Praise for *The General Society of Mechanics and Tradesmen of the City of New York: A History*

The General Society represents all that is right about New York City's building and construction trades. New York's great success is owed in great measure to the men and women who built it and to the organizations that have supported them every step of the way. I am grateful to the GSMT for sharing their inspirational story.

—Richard T. Anderson, President, New York Building Congress, a membership organization comprised of the design, real estate and construction industry

I salute The General Society of the City of New York for documenting its extensive history and building on a distinguished legacy that will benefit generations to come.

—Jay Badame, President and COO of New York, New Jersey and Pennsylvania, Tishman Construction, an AECOM company, the 2008 Craftsmanship Award Recipient

The General Society of Mechanics and Tradesmen and the BTEA have worked in partnership for over a century. A century of experience teaching men and women the knowledge and skills necessary to build the world's greatest skyline and infrastructure that makes New York City the greatest city in the world. That task and partnership will continue into the twenty-first century and beyond. It is a partnership of pride, vision, skill and commitment. Now and forever.

—Louis J. Coletti, President and CEO, BTEA, Building Trades Employers' Association

I commend The General Society for its reverence to the ancestors in this historical monograph, which included my father, who was a Mechanics Institute graduate. My father was one of those who, in 1927, was a proud graduate of the Mechanics Institute.

—Charles DeBenedittis, Senior Managing Director Senior Advisor, Design & Construction Tishman Speyer

The General Society of Mechanics and Tradesmen is a grand place that has bred geniuses who have built New York, and this book gives a dissection of the early members of The Society who did this.

—Charles H. Thornton, PhD, PE, Chairman AECOS/TopDown and the Founding Principal and Chairman of Thornton Tomasetti.

I0806164

THE GENERAL SOCIETY OF MECHANICS & TRADESMEN OF THE CITY OF NEW YORK

A HISTORY

Polly Guérin

THE History PRESS

Published by The History Press
Charleston, SC 29403
www.historypress.net

Unless otherwise indicated, the images in this book are attributed to The General Society of Mechanics and Tradesmen of the City of New York.

First published 2015

Manufactured in the United States

ISBN 978.1.62619.476.2

Library of Congress Control Number: 2015932941

Notice: The information in this book is true and complete to the best of our knowledge. It is offered without guarantee on the part of the author or The History Press. The author and The History Press disclaim all liability in connection with the use of this book.

For the ancestors and the American builders and construction heroes who built this magnificent city and The General Society of Mechanics and Tradesmen of the City of New York that preserves its legacy.

Contents

Author's Note 9
Acknowledgements 13
Introduction 17

1. New York 1785: The Capital City 21
2. The General Committee of Mechanics 35
3. The General Society of Mechanics and Tradesmen of the City of New York 45
4. An Account of Prominent Members 55
5. Members Inspired by The General Society 69
6. Andrew Carnegie and Distinguished Presidents 81
7. Mechanics Hall and Mechanics Bank 89
8. The Mechanics School and The Mechanics Institute 103
9. The Apprentices' Library and The General Society Library 127
10. The Lecture Series and The Small Press Center 149
11. The GSMT Awards 169
12. The General Society Then and Now 181

Epilogue 209
Appendix 1. Landmarks and Resources 213
Appendix 2. Historic English Terminology Used in the Book 215
Bibliography 217
About the Author 221

Author's Note

Throughout this book, The General Society of Mechanics and Tradesmen of the City of New York is referred to interchangeably as "the GSMT," "The General Society" or "The Society."

I have often been asked by friends and inquiring individuals, "How did you decide to write this book about The General Society of Mechanics and Tradesmen of the City of New York?" That's a good question, and it has a simple answer. I enjoy research and historical subjects, and the opportunity to write this book came in a most serendipitous way. The GSMT had invited me to present my previous book, *The Cooper-Hewitt Dynasty of New York*, also published by The History Press, in a PowerPoint presentation and book signing on February 12, 2013.

Since I have been a member of the GSMT's Library for several years, I was very pleased to accept the speaking engagement. The book-signing event was a wonderful experience and well received, but it never entered my mind that the GSMT was more than just the Library where, incidentally, I did research for my previous book.

The opportunity to speak at the GSMT opened my eyes; in fact, everything about The Society was full of significant history. I was intrigued that the 230-year-old organization was still a vital cultural and educational organization located in the very heart of New York City at 20 West Forty-fourth Street.

The Society's stately Library, with its old-fashioned library tables and chairs, the display of historic flags and the multi-level floors of books, seemed a place frozen in time. Yet The Society was in step with everything modern, such as the Mechanics Institute, the oldest tuition-free technical evening school in New York. The Library's spacious venue also plays host to the Mechanics Institute's graduation ceremonies, and many of the GSMT's patriotic celebrations are held there as well as television and film productions and, lately, the screening of an award-winning documentary film.

Shortly after my book-signing lecture, I met with the executive director, Victoria A. Dengel, and asked, "Has anyone ever written a history about the GSMT?" By Ms. Dengel's answer, "no," she also assured me that, while the GSMT had well-documented archives and several pamphlets had been written in early times, no book of any complete scale existed.

I took that "no" to mean that there was an opportunity to write a history of the GSMT being thrust in my direction. So I requested that a meeting be called with the GSMT board, and the subject of writing the book was discussed. There was unanimous agreement that a book would be welcomed, and with the board's sanction, I was given permission to begin research in the archives, which were made readily available to me to work on the project.

I was impressed at every turn to find in The General Society's annals—the historic minutes from 1785 to 1880, a record made by true and honest men who, through their dedication as laborers in the building trades, New York City became the great commercial center of the world.

By their excellent record keeping—as well as historical documents, faded magazine and newspaper articles, and sundry artifacts—the Brothers, as they are called even today, documented their monthly meetings, committee activities, philanthropy and concerns of the organization. They formed a brotherhood organization, founded in the early years of the Republic in 1785, beginning as a group of compatible and focused mechanics, craftsmen and tradesmen.

With the ensuing years, The General Society's astute leaders responded to the critical needs of its members. It gave the city educational advantages with its school and library at a time when no such facilities were open to apprentices. Its crusading benevolence assisted its indigent members, widows and orphans.

Throughout the GSMT's history, patriotism is foremost on the agenda and observed each year, such as the reciting of the Declaration of Independence and events honoring Flag Day and Veterans Day, which continue even today as important historical celebrations in the GSMT's roster of activities.

The American flag is prominent at all of The Society's events, and it also hangs outside on the GSMT building in tandem with The Society's royal blue flag emblazoned with its iconic motto, "By Hammer and Hand All Arts Do Stand." When the dual flags on all the clubhouse buildings are unfurled on West Forty-fourth Street, it reminds me of the American impressionist Childe Hassan's painting, *The Avenue in the Rain* (1917), with vibrant flags unfurled in a patriotic display. This painting incidentally is often chosen by the president to hang in the Oval Office at the White House.

I realize that the GSMT book is the story of old times, old customs and an old building, but it is also an important part of the rich history of New York City that needed to be documented for future generations. Readers will find inspiration in the spirit of the brotherhood that propelled this organization as a viable entity through its 230 years (1785–2015).

New York City may have been built by craftsmen in its formative stages, but today, through the GSMT's tuition-free Mechanics Institute, the people of the city benefit from the graduates trained in diverse high tech programs. It is with considerable pride that the GSMT also continues its vital role in the cultural mecca of New York City through its Labor, Literature and Landmarks and Artisan Lecture Series.

I have endeavored to capture the historical accuracy of the GSMT by turning my archival research into a book with storytelling charm. I hope that my narrative provides page-turner dialogue and vivid imagery about a time when sincere honesty, integrity and brotherhood were the foundation on which New York City was rebuilt after the Revolutionary War.

Today, the GSMT has made new advances in service to the public and continues to uphold its motto "By Hammer and Hand All Arts Do Stand," with activities that impact the business and social life of the city.

In the meantime, worthy reader, I am pleased to present to you a worthy tribute to a worthy organization that made history-changing advances in education and philanthropy and still supports the laboring work of the apprentices, craftsmen and artisans in the building and construction industry.

I trust that through this book you will become better acquainted with The General Society of Mechanics and Tradesmen of the City of New York and its pivotal role as the epicenter that even today continues to serve and improve the quality of life of the people of New York City. The Society does this through its educational, philanthropic and cultural programs, including its tuition-free Mechanics Institute, The General Society Library and its century-old lecture series. The GSMT building itself on West Forty-fourth Street is a pedestrian showstopper as the street-level windows with program information and artisan displays attest to the cultural activities inside.

By the penning of this book, the GSMT's storied past is brought back to life for generations to come. It is time, however, to be reminded that the oldest organization of its kind, The General Society of Mechanics and Tradesmen of the City of New York, was founded in 1785, a mere two years after the official end of the Revolutionary War in 1783. The Society's history, therefore, is unequivocally linked to New York City's history itself.

It is a story of old New York, and it is a story that needed to be told, but it is also the story of the American workers who built this magnificent city. Throughout The General Society's unprecedented record of longevity, the organization has held on to a steady course of continuous operation bringing up the ranks to 2015, its 230th anniversary year.

The General Society of Mechanics and Tradesmen is one of the truly important hallmarks in the history of the City of New York. As such, it is a remarkable achievement, one that supports the interests of America's hero workers who have made, and continue to make, our city the capital of the world.

Acknowledgements

Writing this book would not be possible without the access that I was given to do research in The General Society of Mechanics and Tradesmen's archives, a rare treasure trove of documents, which enabled me to compile the contents of this book.

Much of the credit for documented information, however, goes to the Brothers, the members of the GSMT who conscientiously kept accurate minutes of their meetings in those early years. They also recorded their terms of benevolence and philanthropy and pioneered education for their members' children, which brought about The Society's day school, now The Mechanics Institute. These farsighted Brothers also established the second most viable library in the city. Originally called the Apprentices' Library, later it was renamed, The General Society Library. It is a fully operative site for study and research and available to the general public through membership subscription.

The visionaries of historical significance include Society member (1846) Thomas Earle, vice president (1856) and president of The Society (1857). Earle compiled and recorded the historical *Annals of The General Society of Mechanics and Tradesmen of the City of New York, from 1785 to 1880* with Charles T. Congdon, who is listed as editor. This history, commissioned by The Society, consists primarily of significant documents that figured importantly in my research.

I also wish to extend my gratitude to all the people I worked with or interviewed, without whom writing this book would not have been possible.

From the start, when I brought the book project to her attention, executive director Victoria A. Dengel sanctioned the book, as did the GSMT board at a subsequent meeting. At every step in the book's process, Ms. Dengel was supportive and encouraging. One can surmise that Ms. Dengel, a past president of the GSMT (2011–12) apparently had a profound, innate interest in seeing The Society have an active voice in this book. Perhaps that can be attributed to the fact that she is the daughter of Gerard A. Dengel, past president (1982). In addition, she was a frequent visitor and more or less grew up with the GSMT. Through these early associations, Ms. Dengel developed a strong relationship and interest in The Society. As The Society's current executive director, she continues her earnest reverence for the GSMT's history and focuses on all aspects of the organization's operations and the introduction of innovative programming.

With the board's approval, I went full speed ahead and took the quiet words of wisdom recorded in the archives and turned them into a lively historical discourse. When I would query Ms. Dengel about some written passage, she would say, "Beautiful writing." Those kind words were like music to the author's ears and made the work move along with renewed enthusiasm.

I would be remiss if I did not bring to your attention Angelo A. Vigorito, Society member (2013), the GSMT's archivist, whose knowledge of the legendary organization is so keen that at every turn he was able to unearth documents, clippings, newspapers and annual records of The Society for my perusal and research. In addition, he has a great mind for remembering the important dates of members initiated into The Society and the ability to verify data in the story that I have carved out of the archives and the GSMT's annals.

I am very grateful for Mr. Vigorito's knowledge of the GSMT's history and his scholarly input that made my research possible. In another aspect, although I researched, found and selected images for the book, in a great gesture of thanks, I acknowledge Mr. Vigorito's technical skill as he processed many of those images to accommodate the standards set by the publisher, The History Press. The images in the book, however, unless otherwise noted, are attributed to the GSMT.

With thanks, I must mention John Erroll, curator of the John M. Mossman Lock Collection, for his erudition of the historical aspects of the collection.

I am grateful as well to Joseph G. Reinis, associate curator of the John M. Mossman Lock Collection. Mr. Reinis gave generously of his time and technical skill, working in tandem with Mr. Vigorito.

Then, too, there is Malena Rogers, the GSMT Library's circulation manager, whose knowledge of the Library's multi-floor stacks made book searches her priority. Finding such book relics—books dating back to 1888 and earlier—featured significantly in my research, and at every request, she was a willing support.

As for Karin Taylor, The Society's program director, I do so very much appreciate her input on the Small Press Center of which she was director before its activities were suspended. In her new role as program director of the GSMT, Ms. Taylor brings the same enthusiasm she had for independent publishing and produces the public lectures, which include the Labor, Literature and Landmarks Series as well as the Artisan Lecture Series, both of which are held in The Society's library space. Her input on the historical development of the lecture series is appreciated. In terms of marketing, Ms. Taylor's consummate knowledge in this field greatly augmented my own.

Meg Stanton, who is the registrar of the Mechanics Institute, wears another hat of responsibility and also serves as assistant program manager. She works in tandem with Ms. Taylor in producing the lecture programs.

Then there is Society member (2004) Penny Speckter, the editor of *GSMT News*, the newsletter of The General Society of Mechanics and Tradesmen. She graciously granted me access to her master's thesis, *The Philanthropic Activities of The General Society of Mechanics and Tradesmen of the City of New York, 1785–1885*, copyright 1986. This document was a source of invaluable verification of historical data. Her contributions were sometimes spontaneous. For instance, on any given day she might stop by where I was researching and deliver some bon mot. On one such occasion, she said, "By the way, Polly, did you know about William Webb? He was a Society member and established a school on Long Island. You might want to add something about the Webb Institute." There were other tidbits to follow and a delightful reason to talk with Penny.

I wish to also thank the professionals who contributed their photographs for use in the book, including Society member (2002) Stephen Amiaga, Amiaga Photographers, Incorporated, who also served as GSMT president (2005–06). In addition, I applaud Dan Luhmann and Allan Tannenbaum for their photographic contributions.

I acknowledge the aforementioned individuals with gratitude for their enthusiasm and support during the research for this book. If I have neglected to mention anyone else specifically, I do sincerely apologize for that oversight.

By delving into the silent words of wisdom in the archives, I do truly hope that I have given voice to the ancestors who so valiantly created The General Society of Mechanics and Tradesmen of the City of New York.

Introduction

The history of The General Society of Mechanics and Tradesmen of the City of New York has been a silent record, albeit an excellent one. By summing up the past in this book, I also showcase the future. As The Society moved forward into the twenty-first century, it upheld its fundamental traditions while introducing modern venues and innovative programming.

The founding of the GSMT in 1785 represents over two centuries of excellence, which is unrivaled in its effectiveness as a permanent record in New York City's history. The Society's birth coincidentally also parallels with the very rebirth of New York City itself—a city that struggled back to its former glory after the Revolutionary War. Everything in the GSMT archives was full of material; the footsteps of history were everywhere to be traced and studied to form the basis of this monograph. This book documents the GSMT's storied history and draws from its unpublished archives, minutes and photographs.

At every turn, research was not without its charm, and I was taken up with the ancestors who were true men of integrity, loyalty and patriotic and benevolent character.

Surprisingly little, however, was known about the GSMT in its full glory, and this is regrettable, for the period in which the organization was founded is one of considerable significance in America's labor and commercial history. This tome, however, will shed light on the GSMT's role in supporting the interests of the hero workers who continue to build this magnificent city.

Historical references from the GSMT archives have relevance not only for the workers in the building and construction industry but also for the students attending the Mechanics Institute who will discover in this book the true nature of the GSMT's legacy.

My research, however, was not limited solely to The Society's archives as I also referenced several books about Old New York, which confirmed or augmented historical data in the book. Although numerous images were culled directly from the GSMT archives, I also had the greatest elation when I discovered rare images of historical significance. For instance, I found the image of President George Washington on top of the fort watching the Constitution Parade, the USS *Dunderberg* under construction in William H. Webb's yard and the depiction of New York's great fire of 1776.

You may find the eighteenth-century language of the Brothers like "shipwright" and "millwright" or "cooper" rather strange, but that is how the mechanics of Old New York were known. Definitions of these words are listed at the end of the book.

Please do not be tempted either to assume that the English in which some of the extracted quotes as written are fragmented or incorrect. Indeed that is the way these statements were recorded; the grammar, spelling and punctuation in these extracts are original. I have preserved the integrity of the text exactly as the Brothers of the GSMT wrote it in the annals.

In order to tell this story, it was worthwhile to examine the state of the city after the Revolutionary War because it was during those calamitous times, when the GSMT was founded in 1785, that New York City was being rebuilt. For a few halcyon years, New York was the capital of the United States (1785–90).

However, more significantly, The Society's purpose was not only to rebuild the city and restore commerce but also to improve the condition of its members and their children, who were without social services of any kind. As the book progresses, it unfolds the development of The General Society, which reflects a shift in the character of the organization from a traditional benefit society to an educational organization focusing on its Apprentices' Library, Mechanics Institute and the Lecture Series.

Old-fashioned though they were in the beginning, the men who created the GSMT and those Brothers who followed were, in a manner, "new" in their progressive concepts. For instance, the principles on which they operated serve as a beacon of intelligence and sound principles that apply to organizations even today.

At first, The General Committee of Mechanics, as it was first called, met under humble circumstances in a tavern because at that time there was no other place to hold a meeting. Walter Heyer's tavern, in 1785, was the first site of the organization. There the committee appointed Robert Boyd as its first chairman and adopted its motto "By Hammer and Hand All Arts Do Stand," which featured the "Hammer and Hand" as a symbol of the craftsmen. At a subsequent meeting, the committee was renamed to serve broader purposes and to this day is known as The General Society of Mechanics and Tradesmen of the City of New York.

This book covers a period in which prominent men, some most grievously overlooked in historical reference, were Old New York's builders, mechanics and tradesmen—men like John McComb Jr., who built city hall, and Duncan Phyfe, who was the city's first-ranking cabinetmaker. Industrialist Cornelius H. DeLamater, in collaboration with engineer John Ericsson, built the first ironclad USS *Monitor* and William Colgate, whose company became known as Colgate-Palmolive, figure importantly in the chapter "An Account of Prominent Members."

Industrialists and captains of industry who took up the gauntlet of education were inspired by the GSMT's Mechanics Institute and created schools of their own. Charles Pratt and William Webb and even Peter Cooper, founder of The Cooper Union, credit the GSMT as the inspiration for founding institutes that bear their names.

Brother Andrew Carnegie, a man of large gestures and generosity, figures importantly in this book, as it was through his largesse that improvements were made to the GSMT building. He was not alone as a major donor, other benefactors including Amos F. Eno and numerous other members of The Society also gave generously to support the organization's purpose. Similarly, Society members and corporate sponsors, inspired by this tradition, make financial contributions even today to carry on the work of the GSMT.

The fledgling day school for the members' children evolved into the Mechanics Institute, an institute of national prominence, and takes up an entire chapter in this tome. It is significant that no other institute of its kind exists today in the United States, and the Mechanics Institute continues to provide free tuition in advanced education for individuals working in the building and construction trades.

During the GSMT's quest to find a permanent home to house the Apprentices' Library and the Mechanics Institute, it made an astute real estate investment. "The GSMT's Headquarters" paragraph of this monograph brings the reader to 20 West Forty-fourth Street where the GSMT headquarters maintains its ancestral heritage. Today it moves forward with innovative programming that appeals to modern New Yorkers.

As the epicenter of cultural activities, the GSMT follows in the footsteps of history, taking inspiration from its original lecture series established in 1837. The Society brings luminaries from the building and construction trades to its magnificent library to regale audiences with the latest developments in the Labor, Literature and Landmark Lecture Series. The Artisan Lecture Series introduces culturally rich topics delivered by today's master artisans. Behind-the-scenes venues are also explored in the manufacturing, restoration, preservation and theatrical enterprises.

As the preeminent organization that is so closely aligned with the building and construction trades, the GSMT Craftsmanship and Civic Leadership awards are highly coveted recognition given to members in these trades for their outstanding service and for upholding the standards of integrity and excellence in workmanship. These events get their due discourse in the book, introducing as it does prominent members in these trades that continue to rebuild New York into the Wonder City, a recognition by which it has always been known.

This book is rich in historical narrative that could not possibly thoroughly cover the 230 years of the GSMT's existence in 2015, but it does give a voice to its ancestors and recognizes the skilled workers in the building and construction trades who are an integral part of The Society's and America's futures. This book is a comprehensive overview of an extraordinary organization that has weathered the years, standing forth on terra firma to face the twenty-first century with pride and progress and all its unpredictable glory.

1

New York 1785

THE CAPITAL CITY

New York was a comparatively small city, but it was still a city of great hopes and magnificent prospects. It was predestined to become one of the most important cities of the world, great in mercantile enterprises and the mechanic arts.
—GSMT annals

The founding of The General Society of Mechanics and Tradesmen in 1785 is a noteworthy happening in New York and national commercial history, as it coincided and related to stirring historical events of the time. After the American Revolution, as Manhattan struggled back to power, it was the best of times and the worst of times—a tale of a city in ruin and a tale of a city of great hopes and magnificent prospects.

After seven years of British occupation inflicted on New York, the city incurred immeasurable losses and devastation. Commerce was crippled. The treasury was empty, yet despite these grave setbacks, New York City would prove to be vital to the nation's rebirth.

Some account of New York in those early days, therefore, warrants discourse as epic events set the stage on which the city would begin life anew and settle back into a state of order. It was under these circumstances that The General Society of Mechanics and Tradesmen of the City of New York found its raison d'être and became a major force of dedicated men whose tools and labor would rebuild the city.

It is interesting to note here that the GSMT's history tells the story of its role and New York City's role in shaping the new nation. It also introduces those pioneering men who formed an organization that survives today to carry on its heritage. Subsequent chapters relate the extraordinary story of The General Society of Mechanics and Tradesmen, an organization that even today is the leader in tuition-free education and a cultural giant in the heart of New York City.

NEW YORK UNDER BRITISH OCCUPATION

The circumstance under which New York City valiantly survived the British occupation is a record of resilient American fortitude. When the British showed up in the harbor with the greatest assault force ever assembled, the Americans did not abandon the cause of independence but fought on. It was the only American city that remained under British rule during the entire occupation, and it was heavily damaged by the war.

It is surprising to note that during British occupancy, New York City, for some privileged citizens, was a center of gaiety and fashionable

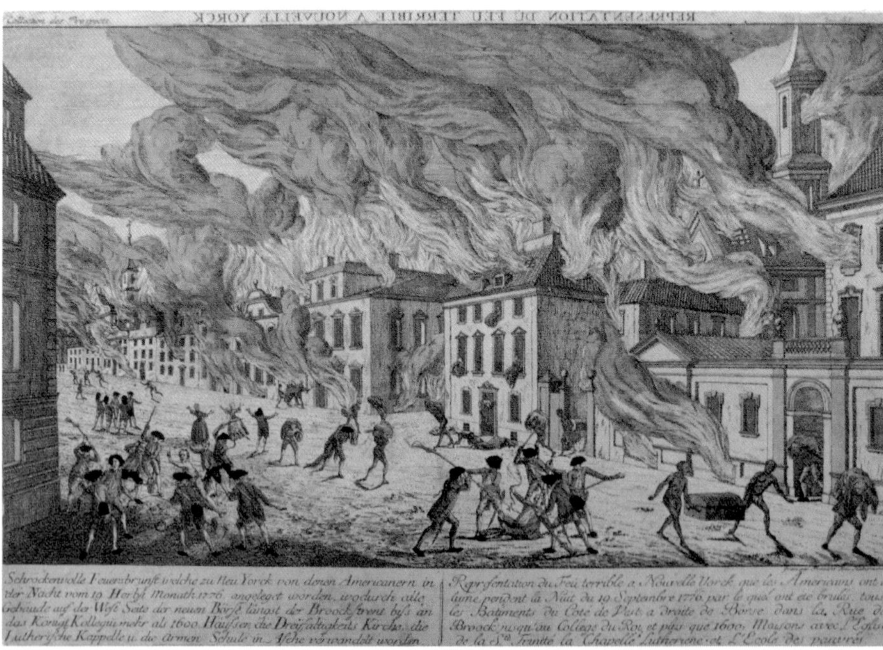

Representation of the terrible fire in New York, September 19, 1776. *By engraver Francois Xaver Habermann. Courtesy of the Amos F. Eno Collection of New York City Views.*

entertainments, but that all ended overnight with the British defeat. However, for the average citizen, the situation was quite different. Houses were plundered, gardens destroyed, public and private buildings used as stables and barracks and garrisons and prisons left in ruin. At this time, New York was a comparatively small city. The inhabitants lived primarily around the southern tip of the present-day metropolis, below Chambers Street, and nearby places like Greenwich Village were merely bucolic suburbs.

A mysterious fire occurred in New York in 1776, when the British controlled most of the city. It was the first great fire in New York. It began in a tavern down by the docks and, driven by the wind, soon swept northward. By dawn, nearly one thousand of the city's buildings had been consumed, now a mere pile of ashes, and long rows of blackened ruins stood in principal streets. Churches bore witness to the destruction; Trinity Church, for one, had to be rebuilt, as did many other historic buildings. Neither side claimed responsibility for the blaze, but the young city that now lay prostrate was by no means dead. Despite its crumbling wharfs and the ruinous state of the city, a new nobler life was to rise from its charred debris.

The city, left in this state of deprivation, was followed by a period of marked depression in commerce and the mechanic industries. Notwithstanding, the able mechanics of Old New York stood in the wings, eager to work, but neither ready employment, or funds to rebuild were available at first.

Under these dire circumstances, the first mayor of New York, James Duane, had the challenging task of restoring property, but the loss of records due to the fire brought personal and business delays of every kind. So great was his service that later, when George Washington was president, he made Duane a U.S. district court judge, and Duane Street in lower Manhattan honors his memory.

Then, too, personal animosities engendered between Whig and Tory, during the years of war, retarded the building of the city. It impacted all work, work that was dependent on the full cooperation of the citizens who had so lately been bitterly opposed in allegiance and purpose. These dissensions, however, were not of long duration, and the folly of regretting the royal government was soon apparent and equally repugnant were the hatred and animosities of war.

The British Troops Depart

After seven years of occupation, the British troops finally departed New York, the last American city to be occupied, on November 25, 1783. With

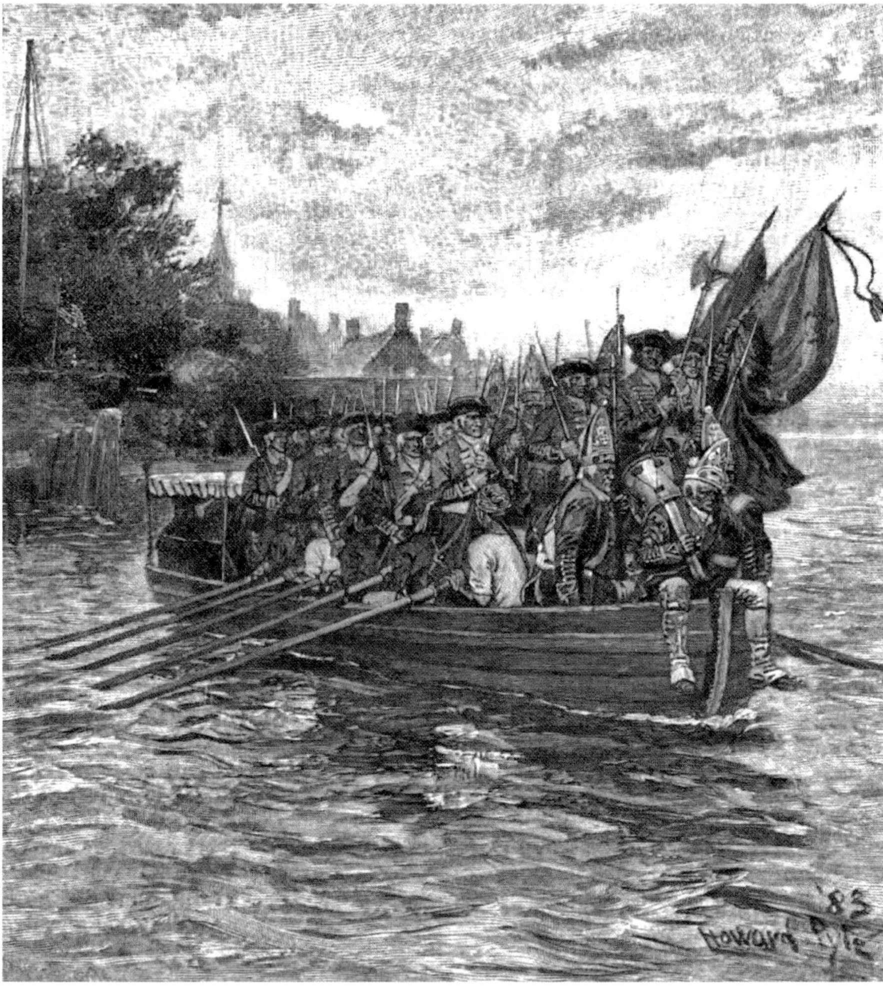

"The Last Boatload of the British Leaving New York" by Howard Pyle, an illustration for Henry P. Johnson's article "Evacuation of New York by the British, 1783." *Courtesy of* Harper's New Monthly Magazine, *November 1883.*

relief and some exultation, the citizens watched as the redcoats rowed off to their ships. The long occupation was over, but the devastated city had much work to reclaim its honorable place in history.

Henceforth, the British departure, celebrated nationally as "Evacuation Day," was observed as a patriotic holiday well into the 1800s. Some of Manhattan's street names, which bore reference to the British occupation, were also changed to reflect the patriotic tenor of the times. For instance,

Crown Street, named in reverence to the British crown, was renamed Liberty Street to honor the emancipation of the American Colonies from British rule.

Of great historical significance is the fact that during these turbulent times, The General Society of Mechanics and Tradesmen of the City of New York was founded in 1785, just two years after the British left the city. This correlation between New York's history and the GSMT's history is a constant reminder that both the city and The Society grew together with magnificent prospects.

George Washington Arrives

On December 4, 1783, after the last British soldiers left America, General George Washington made his ceremonial entry into the city. That night he invited the officers of the Continental army to join him in the long room of Fraunces Tavern, located on the west end of Queen Street, now 54 Pearl Street, where Washington gave his farewell address to his officers.

The first American-born Governor of New York, George Clinton, honored George Washington, the commander in chief, and his officers with an Evacuation Day banquet, saying, "May the remembrance of this day be a lesson to Princes." Amid toasts and tears, Washington embraced each of those present. With melancholy sentiment on both sides, the day was marked by rarely exhibited tenderness among men.

In those early days of New York's rebirth, Congress, desperate for additional space in the war-damaged city, eventually occupied parts of various buildings and also leased rooms in Fraunces Tavern. The Department of Commerce and the Department of War were located there.

The Peace Treaty

The citizens of New York had great cause to celebrate not only their release from British rule but also the circumstances under which New York became the first capital of the nation. There were many patriots responsible for putting New York City in the forefront of news, including John Jay, one of the country's Founding Fathers and president of the Second Continental Congress.

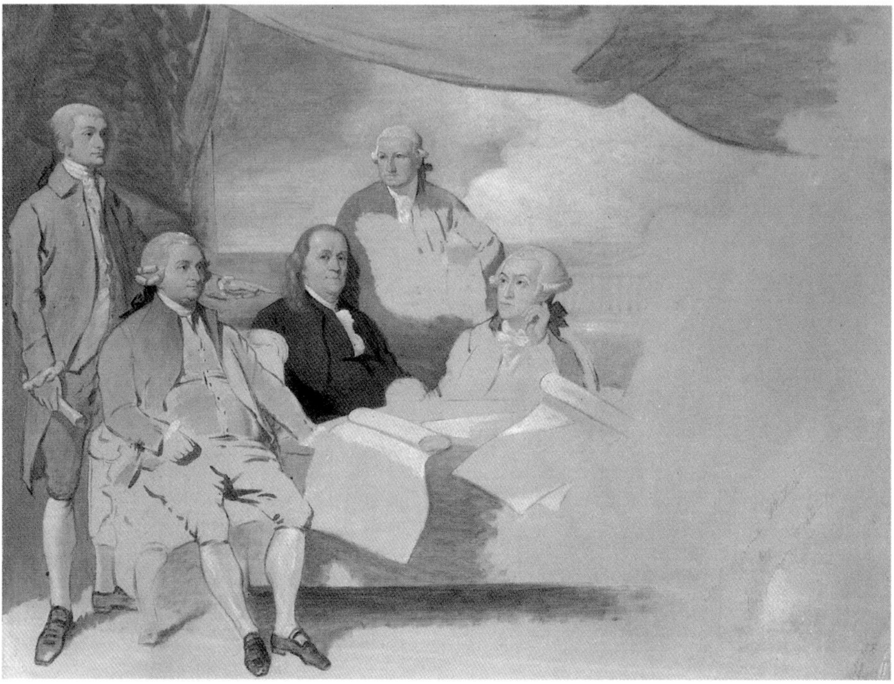

American Peace Commissioners (1782). Copy of Benjamin West's portrait by unknown artist. *Photography by Will Brown. Courtesy of the Diplomatic Reception Rooms, U.S. Department of State, Washington, D.C.*

In his role as statesman, John Jay played a pivotal role in establishing New York as the capital city. However, peace negotiations between Great Britain and the United States took paramount precedence. Jay was a key negotiator of the Treaty of Paris and set the tone for America by refusing to discuss peace until Britain dealt with the United States as an independent nation. "I hope," he wrote, "that America will not find the terms very exceptionable." Only after Prime Minister Lord Shelburne took office in July 1782 did Britain agree to Jay's demand.

Henry Laurens joined the American Peace Commissioner diplomats John Jay, John Adams and Benjamin Franklin two days before the preliminary articles of peace were signed on November 30, 1782. William Temple Franklin was secretary. The peace negotiators proved themselves more than a match for Europe and successfully negotiated and signed the Treaty of Paris on September 3, 1783, which formally ended the Revolutionary War and launched the world's newest nation. (Incidentally, a first class postage stamp was issued in 1983 to commemorate the bicentennial of the Treaty of Paris.)

The War of Independence won and peace now secure, Jay planned to resume law practice in New York City, but with the need for strong leadership in the new nation, Congress elected Jay to the office of Secretary for Foreign Affairs, which put him in charge of diplomatic relations for the United States.

The Capital City

It has been rumored that John Jay's persuasive hand influenced Congress to establish New York as its capital city. Prospects for the city were brightened when Congress conceded and relocated to New York City on January 11, 1785. Thus New York became the official, albeit temporary, capital of the nation from 1785 to 1790. No building was available for even a semblance of a capitol building, and the old city hall was not considered an appropriate place in its present state of disrepair. For a time, venues like Fraunces Tavern became the first office of the government, and officials such as John Jay and Alexander Hamilton met there, as did other prominent and ordinary citizens.

It is significant to note that New York's designation as the capital city, in 1785, coincided with the founding of The General Society of Mechanics and Tradesmen of the City of New York, whose members would eventually meet at Fraunces Tavern as well.

Federal Hall

Several architects submitted plans for remodeling old city hall, but Major Pierre Charles L'Enfant's, a French-born American architect, plan was accepted, and in 1788, the building was enlarged and handsomely furnished, and the name changed to Federal Hall.

The building made a striking appearance. On its façade it was adorned with new crimson-colored balustrades and an ornamented cupola. The building took the fancy of the delegates, and New York was proud of its new public building.

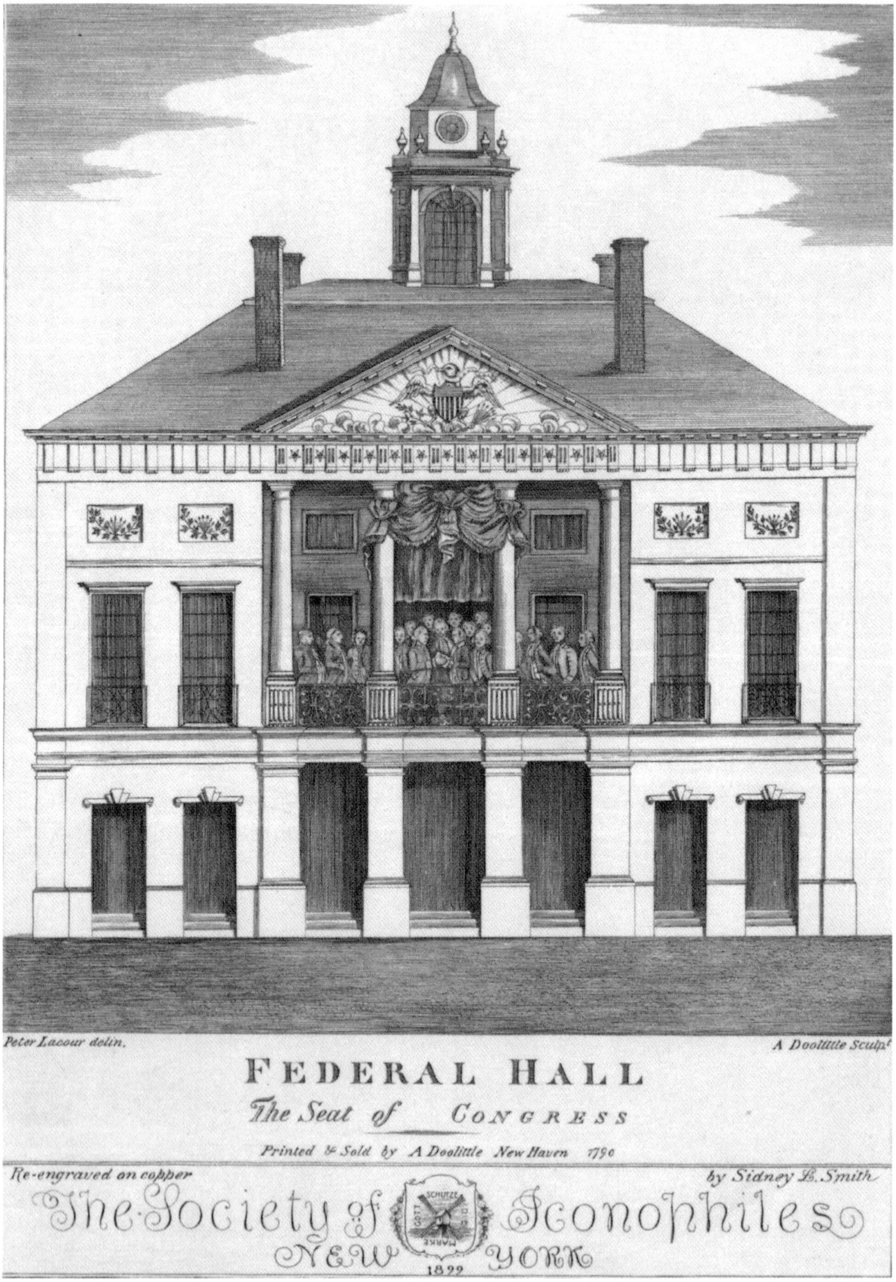

Federal Hall, the first Capitol of the United States and the seat of Congress, 1790. Engraving by Amos Doolittle depicting General George Washington's April 30, 1789 inauguration as the first president. *Courtesy of the Library of Congress.*

The first United States Congress met at Federal Hall on March 4, 1789, to establish a new federal government, and many of the most important legislative actions in the United States occurred there. For instance, on September 25, 1789, the United States' Bill of Rights was proposed in Federal Hall, and in the same year, the Judiciary Act was enacted in the building, which set up the United States federal court system, still in use today.

On April 30, 1789, Federal Hall was the site of an epic event. To commemorate the day, the city was awakened by a gun salute on the Battery, and the inauguration of George Washington, as first president of the United States, took place on the balcony. Chancellor Robert R. Livingston administered the oath of office and then, with unexpected emotion exclaimed, "Long Live George Washington, President of the United States."

Washington, at that time, lived in a four-story house on Cherry Street, on what is now the Lower East Side. On any given day, it was not uncommon for the citizens of Old New York to be greeted by the president. He took his morning ride on horseback along Bowling Green and the Fort Garden, and it was said that he acknowledged the salutes of even the humblest of citizens.

However, Federal Hall's time as the site of the capitol of the new republic was short lived, and the seat of Congress only convened there from 1789 to 1790. Congress met for the last time in Federal Hall on August 12, 1790, and voted to build a permanent capital elsewhere. To this end, Major L'Enfant was selected by President George Washington to design the new capital city on the Potomac River in Washington, D.C. (Today, Federal Hall is a national memorial, a museum commemorating its role as the birthplace of American government and is located in Manhattan's Financial District.)

Restoration of the Shattered City

Meanwhile, the work of restoration of the shattered city, which then occupied lower Manhattan, went on slowly. Business in all occupations kept pace and slowly revived. Citizens with creativity and manpower to build New York up rallied as best they could while great men who had helped to bring about American independence were among the farsighted visionaries who would bring marked improvement and establish the city with such unsurpassed grandeur.

It is noteworthy to mention that, although the mechanics of New York may have been heirs to the Revolution, they would prove their mettle and begin to sow the seeds of brotherhood and cooperation and contemplate organizing for their mutual benefit and commerce. The mechanics, artisans, craftsmen and tradesmen, the founders of The General Society of Mechanics and Tradesmen, would eventually raise the city to such imposing greatness that it would set a new global status for the city.

Despite the fact that New York City was broke, through his instrumentality, Alexander Hamilton successfully created a national financial system and founded the Bank of New York, the first bank in the city and the third in the nation. It commenced modest operations in June 1784 on Pearl Street in the mansion once owned by the merchant William Walton, at that time called the most beautiful home in America.

As the new government organized itself at Federal Hall, New York began to witness a new wave of optimism. There was good reason to be thankful, not only for the deliverance from British rule but also for the restoration of the city. To commemorate the general sentiments of the citizens, President George Washington called for a day of public prayer and thanksgiving and issued the first Thanksgiving Proclamation on October 3, 1789. Little did our forbearers dream that New York was destined to become the first city of the world.

Mechanics and Tradesmen Bond Together

In 1664, the little Dutch settlement of New Amsterdam, at the tip end of Manhattan, had passed into the hands of the English and was renamed New York in honor of James, Duke of York. It is interesting to note that York is a compound of two ancient words, which mean "place at the water," a fitting description for a seaport town.

In those early days of New York, some of the world's greatest shipbuilders—including William H. Webb and Jacob Aaron Westervelt—were located along the shores of New York's East River, particularly on Dry Dock Street. The growth of the city's oceangoing trade and its harbor became the great doorway to commerce and the principal link to trade with China and global markets that assured New York's deliverance from ruin. Merchants from New York and Philadelphia sent the first ship to fly an American flag, the *Empress of China*, to Canton from New York in February 1784.

However, for years during the Revolution, fruitful business had been curtailed in a city of 23,614, whose citizens made a living primarily from its harbor—from imports and exports and the repair and servicing of ships.

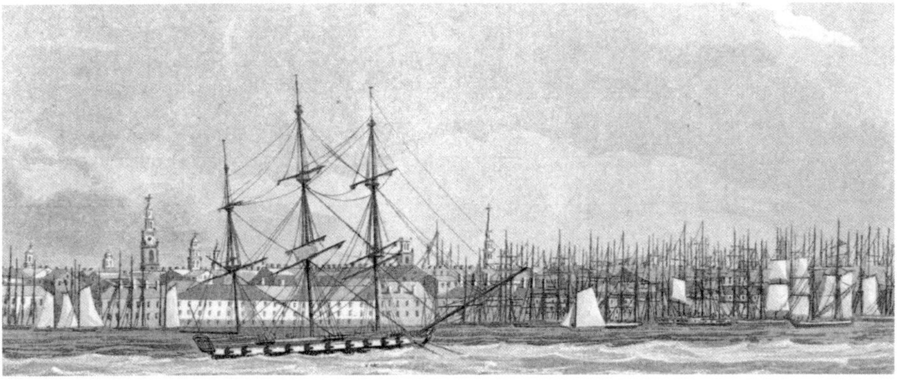

Detail of Old New York's port, the gateway to commerce and global markets. At right, note the ship construction. *Etching by William Guy Wall. Courtesy of the Library of Congress.*

In those desolate years after the Revolution, the average citizen was in crisis. The lack of social services to aid the poor and indigent of the city and public education were nonexistent. The wealthy merchants could employ tutors or send their children to private establishments, albeit modest in scale, but education for the laboring class had not been established.

Responding to these challenging times, the mechanics and tradesmen of the city realized that there was a need to band together, and with the strength of brotherhood and a united purpose, they would promote their mutual benefit, share mechanical knowledge and fulfill their benevolent aspiration to afford relief to deceased or indigent members. It was during these troubled times that the GSMT organization was established. It was destined to wield a powerful influence in the spirit of brotherhood and practical philanthropy into one of the most important institutions in the city of New York.

Concept to Organize

Shortly after the War of Independence, the task of restoring the prosperity to New York City was of paramount concern. During the depression and difficulties of restoration, the idea of forming The Society suggested itself to the mechanics and tradesmen of the city. The concept to organize was for some weeks under discussion, and a few hardworking and skilled craftsmen gathered together with mutual concern, initially meeting at different houses or taverns to discuss the organization's purpose.

You might rightfully wonder that a tavern should be chosen as a meeting place, but taverns in those early years were vital centers of community activity. They were an important link in the new and growing society—places were merchants could conduct business and learn about local news, and tradesmen, craftsmen and artisans could meet in consort and discuss terms of organizing for their mutual benefit.

Most of the businesses in those days were small enterprises. Artisans and tradesmen occupied simple one- and two-story wood houses and lived above their places of business. As competition between like-minded businesses developed, it became necessary for individual shopkeepers to attract trade for their goods and services.

This gave birth to a folk art marketing tool called signboards that hung outside of a shop. These artful attractions were hand printed or carved with designs that identified a business by symbols, among them signs for the tanner, quill manufacturer, coachbuilder or sailmaker. An artisan, for example, advertised all kinds of brasswork with the sign of the brass andiron and candlestick.

However, establishing a brotherhood of men sparked the imagination of the individual craftsmen, artisans and tradesmen. To wit, they continued to meet as a body of like-minded individuals whose mutual interest focused on unification for their common good and benevolent aspirations.

EARLY TERMINOLOGY

It is curious that the word "mechanic" in 1785 had quite a different meaning than it does today. The use of the word "mechanic" in the text may confuse some readers, as they may think that a mechanic applies to an individual who works in the automobile industry. In those early times, the men who would unite in a common bond of brotherly interest were identified by titles that would designate their rank and level of expertise, including the words apprentice, journeyman and tradesman.

For instance, a "mechanic" designation identified a skilled handicraftsman, who owned his own business and worked for himself, directed those under his employ or was a master craftsman who worked for a merchant. He labored in any one of a large number of trades, providing the city with needed services such as blacksmithing, cabinetmaking, shoemaking, breadmaking and quill making, all of which required skill and long periods of training.

An "apprentice" was a beginner, a learner who worked for a master craftsman for a length of time in return for instruction in a trade. For example, a young Scotsman, Duncan Phyfe, who would later become America's most celebrated cabinetmaker, had labored among wood shavings as an apprentice to a coach maker.

A "journeyman" is an individual who has fully served an apprenticeship in a trade or craft and is a qualified worker in another's employ.

An "artisan" describes a skilled craftsman, an individual whose work shows imagination and artistic qualities.

A "tradesman" is defined as an individual who directly deals in, buys or sells raw or manufactured products from a distributive as well as speculative standpoint, also one who employs or directs those employed or may be engaged in related fields of activity. "Related fields of activity" is defined as any part or profession having close contact with production or trade and contributing substantially to their advancement.

In those early days, apprentices and journeymen often received grog from their masters as "coffee breaks" during the day. In later years, that privilege was revoked on the grounds that drinking on the job could be detrimental to production.

Manhattan's rising mechanics used their skills to meet the demands of the burgeoning merchant class. They labored by hand, and the handcrafted artifacts that they created attest to the skill and techniques of a craftsman who, in many cases, invented the very tools needed for their labor.

Pursuit of work in a trade existed within different groups. Most were willing citizens, and some were indentured or forced to work, as slavery still existed. In New York, commercial activity soon grew to include people who had once lived in Africa and Latin America, and this double standard of commerce and diversity has continued in New York's cultural mix.

2
The General Committee of Mechanics

All hail to the hammer; all hail to the hand;
By these all arts flourish; by these all arts stand.
Their Triumph shall spread till this globe they redeem.
And work into fact proudest man's fondest dream.
—*An original motto, GSMT annals*

The vicissitudes of the time required men of great vision to organize for their mutual benefit and commerce. To this end, several like-minded men from diverse trades convened together and began conversing on the subject of bonding together in the spirit of brotherhood.

After meeting on several occasions, the mechanics and tradesmen held their first formal meeting on the evening of November 17, 1785, in Walter Heyer's Tavern, a public house located on Pine Street (then King Street) near Broadway. The organization's purpose was clear from the beginning: it was formed as a response to the disastrous business climate caused by the aftermath of the Revolution, the catastrophic 1776 fire and the end of British occupation two years earlier in 1783.

The meeting was comprised of a committee of twenty-two far-sighted and public-spirited men, including mechanics and tradesmen representing numerous trades—blacksmiths, carpenters, printers, masons, coopers, tailors, hatters, shipwrights and coach makers.

They called themselves the "General Committee of Mechanics" and used their strength in numbers to garner greater respect and recognition

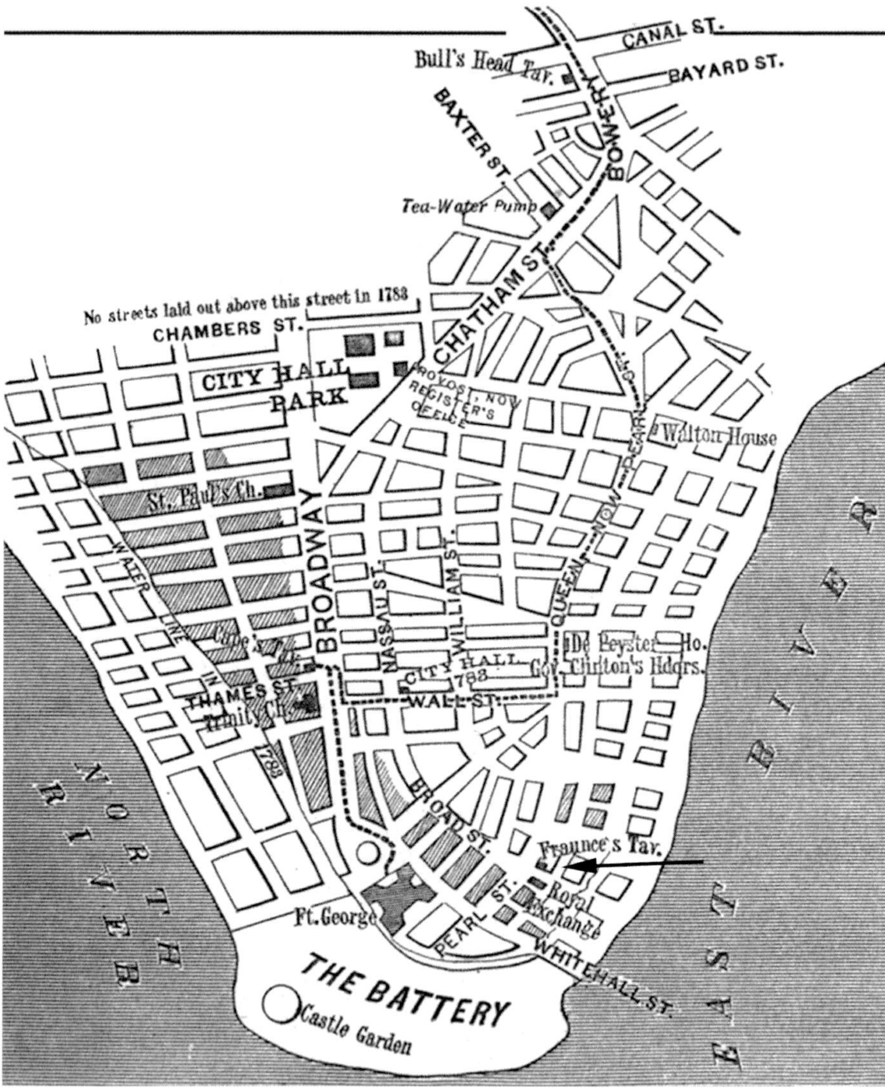

Map showing approximate location of Fraunces Tavern in Old New York. *From the book* "Evacuation Day," 1783 *printed for the author, James Riker in 1883.*

in the city. The committee's purpose was dominated by the spirit of patriotism and devotion to public interests, and the cornerstone of its allegiance was benevolence.

The organization originated under the idea that the burdens of the city could be more easily borne and served in a bond of friendship, protection and mutual benefit. In meetings, members captured the spirit of mutuality and craft pride long ago established in artisan fraternities. High moral standing and exacting regard for honesty and virtue was the capstone of their objective to elevate the city to its former standard of pride. In tandem with these sentiments, the artisans were also anxious to see the new nation claim the power and respect that it deserved.

The General Committee of Mechanics was founded along the lines of practical philanthropy. It instituted one of the first welfare funds and provided assistance for worthy craftsmen and artisans in sickness and distress. It also aided the worn out Brothers and the widows and orphans of members who died without property. The aim and organization of the general committee was typical of a craft benefit society and combined private charity and camaraderie in the bond of brotherhood.

During New York's restoration, so dependent on the laboring class, the committee's prime objective was to encourage the mechanic interests of the city and to encourage the revival of business in the city.

By Hammer and Hand All Arts Do Stand

The General Committee of Mechanics chose a blacksmith, Robert Boyd, as their first chairman, who served in this post in 1785 and 1786. Later he became the organization's first president. At that historic meeting on November 17, 1785, the committee selected the blacksmith's insignia to represent the organization's motto, "By Hammer and Hand All Arts Do Stand," as their own, which asserts the importance of skilled work in the building of the nation and honors the artisans and mechanics who made tools for all trades. According to Alexander James Hill—author of the forthcoming compendium *Measures and Their Makers*, a chronicle of the late seventeenth century through the eighteenth century—"A commonality inherent with these early craftsmen was their ability to individualize their tooling, which they utilized in the manufacture and embellishment of their instruments."

Of the committee men present at the first official meeting were a pewterer, a sail maker, a saddler, a ship joiner, a ship carpenter, a potter, a coach builder, a shoe crimper, a quill manufacturer, a japanner, a solderer, a tallow chandler and a bell hanger, all members of the craftsmanship tradition that identifies with The Society's origin.

Original 22 members, with trade (from the Annals of The General Society of Mechanics and Tradesmen of the City of New York. From 1785 to 1880):

William Allen, Shoemaker
John Anthony, Tanner
John Bailey, Cutler
Henry Buker, Hatter
Robert Boyd, Blacksmith
John Burger, Silversmith
Joseph Jadwin, Cooper
Robert Manley, Coach-maker
Dennis McReady, Tobacconist
Isaac Mead, Carpenter
Edward Meeks, Blacksmith
Anthony Post, Carpenter
John Shepherd, Tailor
William Smith, Block-maker
John Stagg, Mason
George Taylor, Ship-joiner
Andrew Thompson, Jr., Bricklayer
Hugh Walsh, Tallow Chandler
Henry Wolf, Comb-maker
Isaiah Wool, Stonecutter
Ebenezer Young, Ship-carpenter
John Young, Saddler

The original twenty-two members of The General Society of Mechanics and Tradesmen of the City of New York, 1785.

At the first meeting of The General Committee of Mechanics, it was resolved that a book should be provided for the purpose of inserting the minutes and a chest be provided for the purpose of depositing funds of The Society. At some point, the committee established an initiation fee of sixteen shillings and a monthly sum for dues thereafter, which would substantially fill their coffers for philanthropic work. These funds regularly received into the treasury were safely invested, while members to whom loans were made felt that they had a good and trustworthy credit.

Meanwhile, the Committee was pressing on and establishing its status in the city as a fraternal organization that would figure importantly in the education and cultural development of New York City. Society member (1854) Wade B. Worrall—second vice president (1865), vice president (1866) and president (1867)—had the following to say in his inaugural address:

How little dreamed the men of 1785 that the small humble affair of their creation, a simple society of mechanics organized in a small community banded together for mutual relief in case of sickness. How little dreamed they, in their small beginning, of the superstructure that was to rise on the foundation that their hands were laying.

At the general meeting, held again at Walter Heyer's house on the evening of November 24, 1785, it was resolved that three hundred copies of the organization's constitution and bylaws be printed for use of the different branches. It was signed by all present except one—Isaac Mead, who desired time for further deliberation. As recorded in the minutes:

> *When the twenty-two men first agreed to aid and assist one another, little did they imagine to what extent their initial efforts would benefit the mechanics of New York City in years to come.*

Membership Requirements

The general committee was composed of several different trades, and each trade had a separate organization, which was considered a branch of the committee. At the same place, Heyers Tavern, on Wednesday evening, December 7, a meeting was held for the purpose of receiving subscriptions, and new members were welcomed into the fold.

Membership was limited to craftsmen who were at least twenty-one years of age. The bylaws also stipulated that a candidate be proposed by two to four members who would attest to his "industry, honesty and sobriety." There was an initiation fee of five dollars, and monthly dues were twelve and a half cents. The membership committee gave careful consideration to the qualifications of candidates for membership and exercised careful judgment in the selection of members from those whose names were proposed for membership:

> *Applicants must be a citizen of the United States and a mechanic or tradesman. The name of every person applying for membership, together with his art or trade, present occupation, place of business and residence, shall be referred by The Society to the Membership Committee, accompanied by the recommendation of at least four members who in their own knowledge do in writing vouch for the character and qualifications of the candidate. If the Membership Committee finds him to be eligible, the applicant is then reported to The Society to ballot on the question of admission to membership.*

Few things display better the dignity and pride of The Society than Robert Boyd's presidential address at the initiation of new members, who henceforth would be called "Brothers."

> *You have been regularly admitted a member of this Society, which mark of favor with which our institution has honored you is a happy result of those flattering recommendations which we have received of your character. Let sobriety, industry, integrity, and uprightness of heart continue to be the ornaments of your name. We now hail you brother! a delightful union, where the bond of friendship is benevolence.*

At this time, the citizens of New York were inspired with confidence in the future of the neophyte republic. Political representation opened up as early as May 1785 and included George III's reception of John Adams, the representative of the United States at the court of Great Britain. Then the arrival of John Temple, the first counsel-general from Great Britain, on November 23 of the same year, gave evidence that the political climate was changing with mutual respect between the United States and Britain.

PAGEANTRY AND THE CONSTITUTION

In New York, most citizens favored the constitution calling for a strong central government. There was, however, interminable delay in adopting the Constitution of the United States, and New York was among the laggards. Meanwhile, the new federal constitution was ratified by state after state, yet New York dragged its undecided heels. By June 21, 1788, New York had either to unite with the majority or withdraw from the Union.

While the sessions of the assembly had met but seemed unable to reach a conclusion, the people of New York had grown restless waiting for action and proceeded to carry out a plan to impress upon the decision-makers to go forward. To this end, the city was ready to accept the developing concepts of federalism and staged a great parade to exhibit the pride and products of the city's mechanics and tradesmen. The metropolis thought that such an exhibition of the popular feeling would materially influence the obstinate body at Poughkeepsie and bring matters to a crisis.

A committee was set up under the supervision of Major L'Enfant (the architect of Federal Hall), who organized ten divisions for a parade. Martha J. Lamb's book *History of the City of New York* (1880) describes the magnificent scale of the proceedings known as the "Constitution Parade":

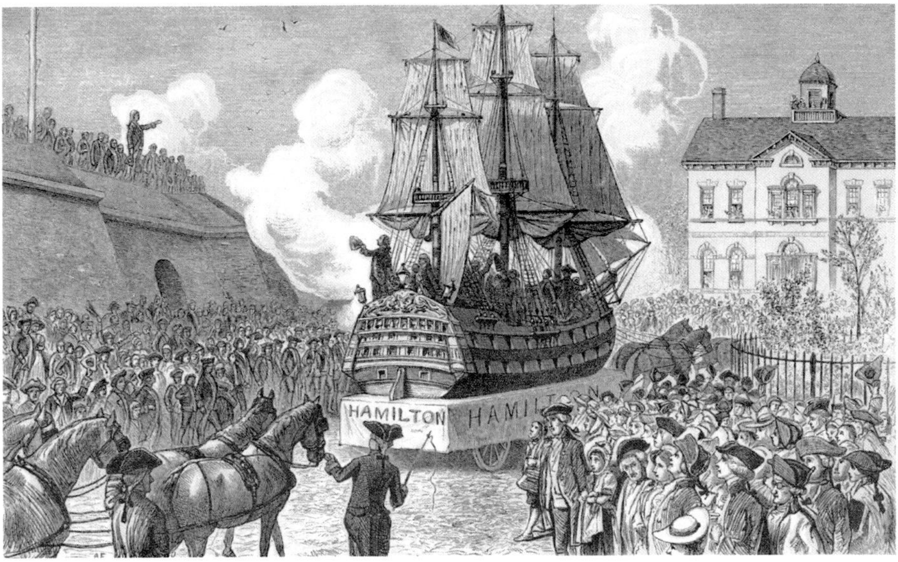

George Washington and members of Congress standing on the fort salute the *Hamilton* ship while watching the Constitution Parade, 1788. *From* History of the City of New York *by Martha J. Lamb.*

The morning July 23, 1788 was ushered in with a salute of thirteen guns from the Federal ship Hamilton, *moored off Bowling Green. The vessel had been built for the occasion and presented by the ship carpenters. It was equipped as a frigate of thirty-two guns, twenty-seven foot keel and ten beams, with everything complete in proportion, both in hull and rigging, and was manned with upwards of thirty sailors and a full complement of officers, under command of the veteran James Nicholson. By far the most imposing part of the gorgeous pageant was the Federal ship, heading the seventh division, with Hamilton's name emblazoned upon each side of it and it moved slowly through the streets drawn by ten magnificent horses.*

When opposite Bowling Green, President George Washington and members of Congress were discovered standing upon the fort, the ship was instantly brought to and fired a salute of thirteen guns, followed by three cheers, which were returned by the Congressional dignitaries. Five thousand men and boys representing some sixty odd trades all showed up all dressed to represent their various trades with colorful floats and banners proclaiming the happiness and prosperity that would follow from a stronger national union. The procession was formed upon a scale of vast magnitude, and it be the first of the kind in New York—or in America—which lathering since has excelled in magnificence of design or splendor of effect, a brief outline of its principal features will vividly illustrate the spirit of the age.

The decorations and costumes of the societies were of the greatest variety and significance. Most interesting, as the parade progressed, were the tradesmen, craftsmen and artisans who demonstrated their skills working aboard horse-drawn carts. Numerous trades marched in formation, thus proclaiming their allegiance to the new government.

The coopers in white leather aprons and oak branches in their hats marched in formation and upon a cart were coopers at work, constructing an iron-bound keg which bore the inscription "New Constitution."

Another division commenced with the carpenters, who numbered altogether upwards of two hundred; each carried a rule in his hand, and a scale and dividers hung from his neck. The cabinetmakers illustrated their craft with a cart drawn by four magnificent white bay horses, upon which a table and a cradle were completed during the march and the chair-makers, sixty or more, marched valiantly with the motto upon their standard:

"The Federal States in union bound,
O'er all the world our chairs are found"

The blacksmiths and nailers, numbering 120, began and completed an anchor on their stage during the march, while their motto floated in the breeze:

Forge me strong, finish me neat
I soon shall moor a Federal fleet."

The stonemasons displayed the Temple of Fame supported by thirteen pillars, ten finished and three unfinished, with the inscription:

The foundation is firm, the materials are good,
Each pillar's cemented with patriot's blood.

The most striking contingent was the eighth division, in which some 300 cartmen marched in their white frocks. They carried a banner upon which was inscribed:

Behold the Federal ship of fame
The Hamilton *we call her name*

To every Craft she gives employ
Sure cartmen have their share of joy.

The image of Alexander Hamilton was carried aloft on banners in every segment of the procession, the Constitution in his right hand and the Confederation in his left. He had to all appearances turned the scale for the Union, and fame was indeed crowning him with well-earned and enduring laurels. A few carvers and engravers; coach and harness makers also followed on a large stage with men at work. At the final flanks walked the clergymen, physicians, scholars, gentlemen preceded by a blue flag with the motto, "United we stand, divided we fall." The novel procession moved along New York's chief streets, watched by multitudes even to the housetops, no sounds heard save the horses' hoofs, carriage wheels, the necessary salutes and signals, and the crowd-pleaser jubilant reaction.

This magnificent parade had the desired effect. Three days later, on July 26, New York became the eleventh state to ratify the Constitution that brought about a loosely allied confederation of states under a strong central government that derived its power from the people. By September 13, 1789,

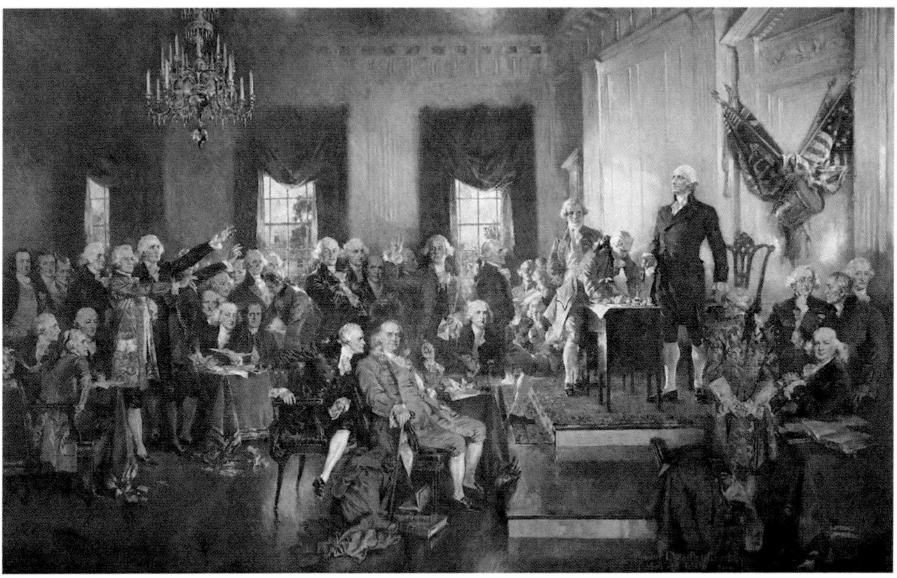

Signing of the Constitution. George Washington, foreground, Benjamin Franklin, Alexander Hamilton and James Madison at Independence Hall, Philadelphia, 1787. *By Howard Chandler Christy. Courtesy of the Architect of the Capitol, Washington, D.C.*

the adoption of the Constitution was publically declared, and New York was selected as the seat of government. The general excitement was so great that bells peeled one long, loud cry of joy; and from the fort and the federal ship *Hamilton*, the discharge of artillery was deafening.

IN A MANNER OF DRESS

The patriotic citizens of New York viewing the parade on that auspicious day turned out in their Sunday finest. Following the period after the American Revolution, most Americans rejected English designs and turned to the French for inspiration. In those early years, the merchant class and tradesmen were stylishly dressed, sporting a cutaway tailored coat, a waist-length waistcoat and knee-length breeches. The adornment of wigs was de rigueur, and a tricorne hat was a proper accoutrement.

The fashions worn by society's belles were also influenced by French court dress: even their children's clothing bore the Gallic influence. A fanciful hat trimmed with ostrich feathers accompanied a gown with a fitted bodice and full skirt that parted at front to reveal an inner layer or petticoat.

As for the rest of the population on the day of the great parade, the average citizen simply emulated society by wearing adaptations of the prevailing fashion.

3

The General Society of Mechanics and Tradesmen of the City of New York

The Mechanics Committee takes the name by which it is still known, two hundred and thirty years after its founding; The General Society of Mechanics and Tradesmen of the City of New York.
—GSMT annals

In 1786, less than a year after The Society's founding, the organization's membership had increased substantially and discussion was underway that would make a significant name change. In order to encompass a greater representation of its members, the brotherhood decided to improve on its General Committee of Mechanics title.

In the New York City directory of 1786, the mechanics committee had already taken the name by which it is still known 230 years after its founding, The General Society of Mechanics and Tradesmen of the City of New York. In its abbreviated form, it is referred to interchangeably as "The Society," "The General Society" or the "GSMT."

Forty-four members are listed in the directory in which the original twenty-two members are included. The Society was making significant progress, and through a ballot poll at the first annual meeting, dated January 2, 1786, the following three founding members were elected: Robert Boyd, a blacksmith, was officially reelected chairman; John Burger, a silversmith, was deputy chairman; and Hugh Walsh, a tallow chandler, was treasurer. James Bingham was appointed secretary for the ensuing year. At this first annual meeting of The Society, the following curious resolution was adopted:

Resolved, that forever hereafter, on the annual meeting of The Society, no person be permitted to smoke the tobacco-pipe during the time of business.

This non-smoking rule still prevails today at The General Society's headquarters.

CERTIFICATE OF MEMBERSHIP

At a subsequent meeting, held August 2, 1786, a certificate of membership was under discussion. A committee was appointed to engage the artist Abraham Godwin to design and engrave a plate for a certificate of membership. The design and price having been previously agreed on, Godwin charged eleven guineas for this work, and the printed certificates went into circulation.

Although Godwin was not a member of The Society, he was nonetheless inspired by the mechanics, craftsmen and tradesmen; recognized their role in the economic enrichment of the city; and portrayed them with appropriate motifs. The framework design centers on The Society's iconic motto "By Hammer and Hand All Arts Do Stand" at the top center, with a muscular arm and hand firmly gripping a hammer surrounded by the laurel wreath that represented the members' feelings of unity, pride and ambition with skilled labor and hard work building civilization.

The certificate places in its four corners images illustrating The Society's role in the new nation. America, the land of opportunity, is depicted in the upper-left drawing where an Indian and two women stand, one holding a honeycomb. In the lower-left corner, an American is shown plowing the land and building a mill. On the top right, the fraternal aid given to fellow mechanics by their Brothers is illustrated by an artisan offering financial aid to a craftsman's indigent widow. She appears as if waiting, with her arm outstretched as an aid-bearing member of The Society arrives.

Hardworking artisans on the certificate include the turner, blacksmith, housewright and shipwright. The common Masonic symbols, in the lower right, represented some of The Society's members who were Masons. This engraved metal plate remains in the archives of The Society today.

A seal was also struck at the same time for use in connection with the membership certificate to engender The Society's charitable assistance and egalitarianism. It features the image of "Equality, Viz," an aged woman holding a pair of swallows in one hand and a nest of swallows in the other

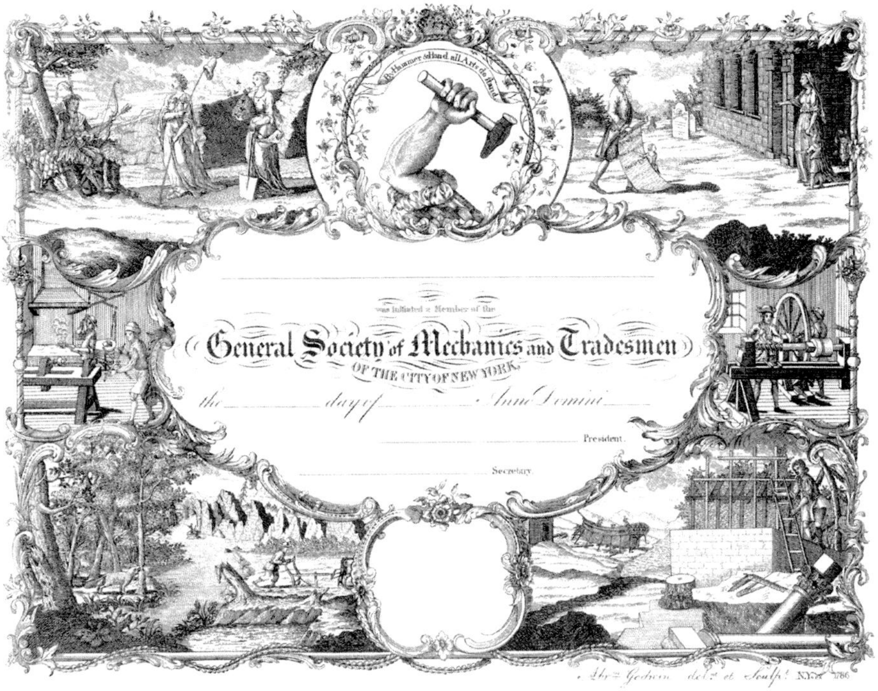

The General Society membership certificate by Abraham Goodwin, artist and engraver, 1786.

fed by the old one. The seal is centered below on the membership certificate. Godwin's bill for engraving the certificate and seal amounted to $59.33 in today's currency. It was ordered and paid for by The Society.

A NEW SLATE

A changing of the officers of The General Society was determined each year by secret ballot, and by 1787, a new chairman, Society member (1786) John Campbell, was voted into the ranks. His appointment is of significant historical interest. Campbell, a former general in the Continental army, is the grandfather of Peter Cooper, the founder of The Cooper Union. The secretary at that time was Society member (1791) John McComb Jr., soon to be the architect of the new city hall. Peter Cooper became a member later and joined The Society in 1837. The Society would eventually open a free school and library. However, before those developments, tariff protection against imported goods was of paramount concern.

Tariff Protection

The General Society was one of the country's most prominent trade societies and promoted not only the social and educational interests of the members but their political concerns as well. The organization made itself a force felt nationally and fought for protective tariffs. All through the years, The Society was corresponding with like-minded groups of businessmen and manufacturers in other leading cities to petition the state legislature in the interests of industrial progress. On November 17, 1788, members of the organization, acting in consort with manufacturers, sent out a letter to mechanics and tradesmen elsewhere. The *New York Journal-American* reported on this development:

> *It joined cause with them against "a pernicious system of commerce which was introduced into our country and which has opposed the extension and improvement of American manufacturers."*
>
> *The article continued:*
>
> *"When our views, like our interests, are combined and centered, our petitions to the federal legislature will assume the tone and complexion of the public wishes and will have a proportionate weight and influence."*

Tradesmen, including those of The General Society, supported the federalists who favored tariff protection during the ratification of the United States Constitution in 1789. Although The Society was a prime mover in this action, other like-minded organizations contributed support. In a letter from the Associated Mechanics of Boston, dated September 20, 1785, and read at a meeting of the mechanics committee on September 26, 1788, a petition regarding the importation of certain articles from Europe stated:

> *We therefore apply to you gentlemen to lend us your assistance, and like a band of brothers, where interests are wisdom dictate. As we hope to experience the good effects of the late arts of our general court, we would*

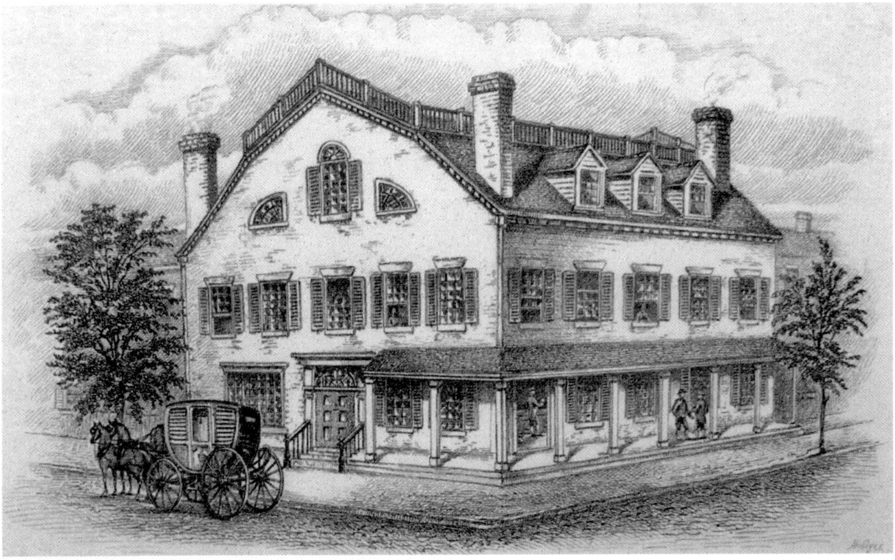

Fraunces Tavern where the GSMT's members met in 1788. *Artist rendering by Samuel Hollyer, author of the book* Old New York Views.

> *recommend a petition for a similar purpose to your legislature and from the known disposition to promote the welfare of America, we doubt not some plan will be devised by your general assembly to prevent the importation of such species of articles as are commonly manufactured in America. We need not urge the necessity of some measures being immediately taken by the whole confederacy.*

Meeting at Samuel Fraunces's Tavern, on Tuesday evening, November 18, 1788, an extra session was called with John Campbell, the reelected chairman presiding. The General Society annals documented the proceedings:

> *Every zealous and intelligent friend to the prosperity of this country must view with peculiar regret the impediments with which foreign importation has embarrassed the infant arts in America. We are certain that they are not only highly unfavorable to every mechanical improvement but that they nourish a spirit of dependence which tends in some degree to defeat the purposes of our late Revolution and tarnish the luster of our character. In order to support and improve the union and harmony of the American manufacturers, and to render as systematic and uniform as possible the design for the common benefit, we perfectly*

> *consort with you on the propriety of establishing a reciprocal and unreserved communication.*

At a meeting of the General Committee of Mechanics on Wednesday evening, April 1, 1789, it was resolved to petition the federal legislature in favor of American manufacturers and "appointed as the subscribers of a committee to correspond with our brethren within the Union." In a letter from Baltimore:

> *The present melancholy state of this country; the number of the poor increasing for want of employment, foreign debts accumulating, houses and lands depreciating in value, trade and manufactures expiring—this being a faint skeleton of the gloomy picture this country exhibits, it is the Supreme Legislature of the United States, as guardians of the whole empire, that every year is now directed from their united wisdom, their patriotism, their ardent love of their country, your petitioners expect to derive that aid and assistance which alone can dissipate their just apprehension and animate them with hopes of success in future, by imposing on all foreign articles, which can be made in America, such duties as will give a just and decided preference to our laborers, and thereby discountenancing the trade that tends so materially to injure them and impoverish the country, and which may also in their consequences contribute to the discharge of the national debt and the due support of government.*

So severe were the grave concerns of the mechanics committee that the legislators of the State of New York, in senate and assembly, were duly influenced and convened to pass the necessary legislation to impound duties on imports. A *Journal-American* feature further reported:

> *The weight and influence brought about legislation that sponsored new industries in the infant United States, protected them and abetted the extension of trade. It has been suggested by more than one historian that it was the weight and influence of the organized mechanics and manufacturers that brought about the ratification of the new U.S. Constitution in the face of opposition.*

It is interesting to note how early in the history of the country the discussions of tariff and free trade, protection and revenues began, topics that still occupy the attention of our citizens today.

Brotherhood Celebrated

The activities of The General Society continued to focus on the sound principles of brotherhood. At the annual meeting, held on January 6, 1789, Society member (1785) Anthony Post, one of the original twenty-two members, was elected chairman. The following Brothers were elected officers for the ensuing year:

> *Society members (1786) William J. Ellsworth, Treasurer and (1791) John McComb, Jr., Secretary and (1787) James Bramble was elected Deputy Chairman. After the election, the members dined together, and during the conviviality's the following characteristic toasts were enthusiastically received:*
>
> *"To the United States of America."*
> *"His Excellency, George Washington."*
> *"The Land we live in: may it ever be free!"*
> *"May the mechanics of the United States discover ingenuity in their profession and honor in their dealings!"*
> *"May our effort to do good be crowned with success!"*
> *"May we live and our enemies know it!"*

This celebration proved so pleasant that at each return of the anniversary, the dinner was repeated, thus cementing the bonds of friendship that bound the members together.

The Original Charter of The Society

Quickly growing in membership and influence, in February 1791, The Society considered the matter of applying for a charter of incorporation. There was great discussion and deliberation, which extended to an annual meeting held on January 3, 1792, when new officers were in place. Past chairman Robert Boyd reassumed his leadership position and was elected the first president of The Society, the title of chairman having been changed to that of president in 1791. Society member (1785) Anthony Post, the former chairman, was elected vice president. Three members who had joined The Society in 1791 were elected as follows: Daniel

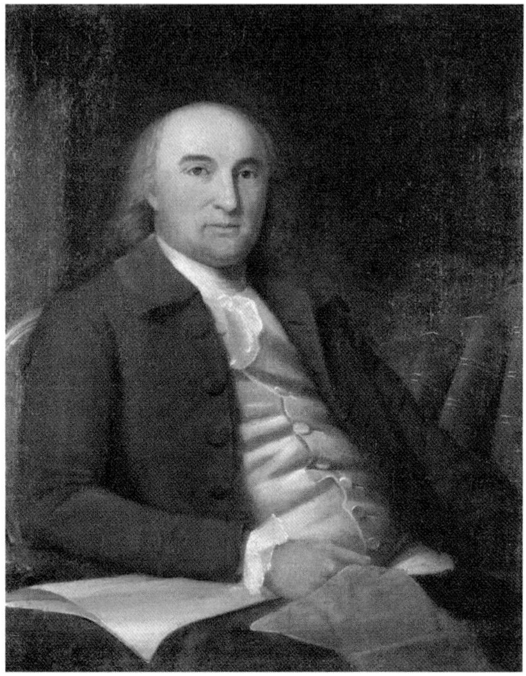

Robert Boyd, GSMT chairman (1785) and its first president in 1792, the year of The Society's charter. *Courtesy of the Brooklyn Museum Open Collection.*

Hitchcock, second vice president; Thomas Le Foy, treasurer; and William J. Ellsworth, secretary.

With President Robert Boyd steering the discussion, and after deliberation and consent of the members, an application to incorporate The Society of Mechanics and Tradesmen of the City of New York, for charitable purposes, passed on March 14, 1792:

> *Whereas, Robert Boyd and others, Mechanics and Tradesmen of the City of New York, associated as a Society under the style of "The General Society of Mechanics and Tradesmen of the City of New York," for the laudable purpose of protecting and supporting each of their brethren as by sickness or accident may stand in need of assistance, and for the relief of the widows and orphans of those who may die, leaving little or no property for their support, by their petition, presented to the Legislature, have prayed to be incorporated to enable them to beneficently carry into effect their charitable intentions. The charter further provided, that the lands, tenements, and hereditaments, which it shall be lawful for the said corporation to hold, shall be only such as shall be requisite for the purpose of erecting a house or hall thereon, or such as shall have been bona fide mortgaged to it by way of security, or conveyed to it in satisfaction of debts previously contracted in the course of its business, or purchased at sales on judgments which shall be obtained by debt.*

SPIRIT OF PATRIOTISM

The General Society of Mechanics and Tradesmen of the City of New York was represented by men of integrity and sound business practices and held a position of prominence in New York. In a show of solidarity and respect, The Society participated in various demonstrations of allegiance to the men who had helped to rebuild the city.

When the eminent financier and statesman Alexander Hamilton, who had practically established the credit of the country, was killed in the duel with Aaron Burr in 1804, the members of The Society met and agreed to wear mourning for six weeks. With similar reverence to the death of George Washington in 1799, which The Society deemed a "melancholy occasion," the chairs of the president and vice president of The Society were hung with black drape and the painting of Washington above the president's chair was festooned with black drape and tied with ribbon.

Throughout its career, the spirit of the loftiest patriotism and devotion to public service has dominated The Society. It has participated on many occasions in notable public ceremonies, such as the peace celebration

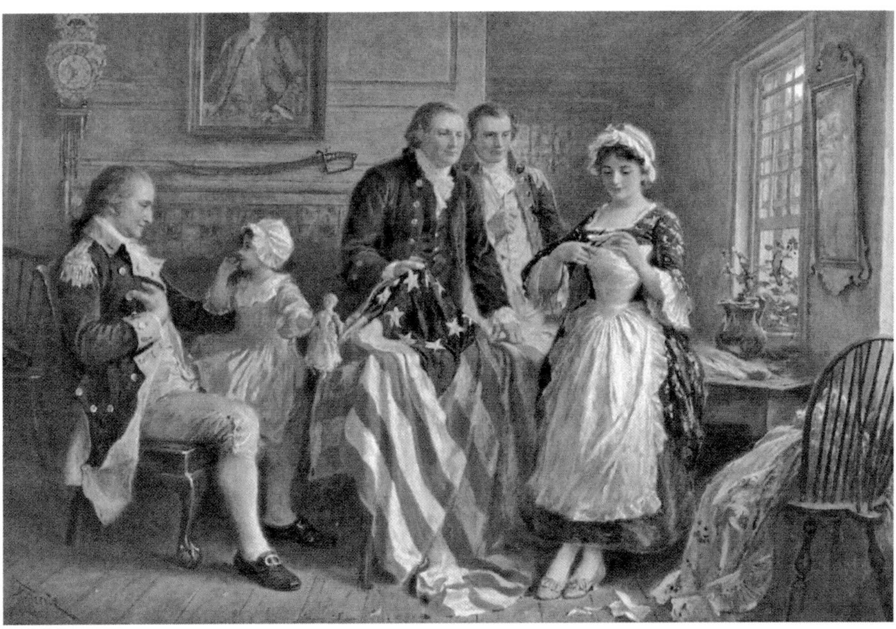

Betsy Ross sewing the stars on the American flag in 1777. A 1920 depiction by Jean Leon Gerome Ferris showing George Washington (seated left), Robert Morris and George Ross. *Courtesy of the Library of Congress.*

in 1814, the Centennial of Evacuation Day in 1883 and the Washington Centennial in 1889.

Of all the patriotic events, the GSMT's annual Independence Day celebration is a custom that has continued up to the present time. Reading of the Declaration of Independence is always part of the ceremony.

The Society is also represented in the annual Flag Day tradition, which honors the 1777 adoption of the American flag by the Second Continental Congress. The distinctive stars and stripes serve as a stirring reminder of the courageous patriots from all walks of life who joined together to build the new nation.

The GSMT continues this tradition and is represented in New York City's Flag Day parade, with General Society members marching fourth in line and alternating the honor of carrying The Society's colors.

The General Society of Mechanics and Tradesmen stands as a record that honors the history of the United States, as well as the history of New York, and glorifies the role of artisans, craftsmen, mechanics and tradesmen who figured importantly in New York City's cultural and economic development.

The Society is best known not only for its solvency in business but most significantly, at this time, as a brotherhood organization that took care of its own members with trailblazing philanthropy.

4
An Account of Prominent Members

In the first fifty years, more than 664 men, from 102 trades and crafts who called their skills "mechanical" or "handmade" arts joined The General Society.
—GSMT annals

The General Society's Membership Registry book not only provides the names of its members and the date of their membership, but it also offers a fascinating look at the trades that were represented by the diverse industry of its members.

In 1824, for example, occupations/trades included bookbinders, carvers, sail makers, painters, a hatter, cabinetmakers, a ship carpenter, a butcher, a brass founder, turners, a hair dresser, shipwrights, a Morocco dresser, a gilder, a carpenter, a mason, a taylor [*sic*], a watch maker, a leather dress, shoemakers, a dyer, a printer, saddlers, chair makers, a stone cutter and a boat builder. These men, through physical labor, helped to shape the entrepreneurial spirit that has always been identified with New York.

The Brothers of The General Society who are represented in the GSMT annals for their contribution to the city's rebirth, are too numerous to mention, but some of the prominent men presented herewith will resonate with readers' historical interest.

JOHN MCCOMB JR.

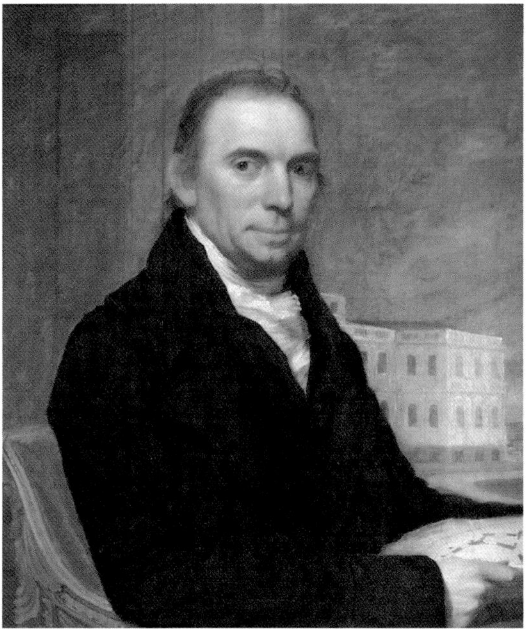

John McComb Jr., architect of the new city hall. *Painting by Samuel L. Waldo. Photograph by Glenn Castellano. From the collection of the City of New York. Courtesy of the Design Commission.*

One of The Society's early members was the Scots-Irish American architect John McComb Jr. (1763–1853), who is best known for his work on the new city hall, built from 1803 to 1812. The GSMT played an important role in McComb's life. He continued to be involved in The Society's development throughout his long architectural career, which contributed to his reputation as the master-builder of New York.

McComb's involvement in The General Society attests to his loyalty to the organization. In the GSMT's 1916 manual, McComb is listed as secretary, a position that he held from 1787 until 1791 while he was not yet a member of The Society. After his tenure as secretary, McComb became a Society member (1791) and was elected second vice president (1816), vice president (1817) and president of The Society (1818).

Throughout his career, McComb was involved with major architectural projects and city planning. He was also the designer of Alexander Hamilton's home known as the Grange, which was named after Hamilton's ancestral home in Scotland. Adding to McComb's credentials, he was a city surveyor (1813 to 1821) and a street commissioner.

It is interesting to note that McComb's father—John McComb Sr., the esteemed mason and builder—had joined The Society in 1786, one year after The Society's founding. He built a number of churches, and according to the records of St. Peter's accounts book, John McComb Sr. was the mason of the first Roman Catholic church in New York. McComb

Jr. became his father's apprentice between 1783 and 1786 and, with his father, built many firehouses around New York after the Revolution. Both were also volunteer firemen.

The New City Hall

1802

Visitors to New York City would have been impressed at the speed with which the city was being rebuilt. The New York City Council discussed plans to erect a new city hall to replace Federal Hall, which had served as the first capitol building of the United States from 1789 until 1790. However, the financial strains of the Revolutionary War delayed the building's progress.

The council persevered in its plan and chose a site at the old common, in lower Manhattan between Broadway, Park Row and Chambers Street, now City Hall Park. The council wanted a new building to house the city hall and courtrooms. To this end, in 1802, the council announced a competition and advertised in the *American Citizen* and *General Advertiser* calling for proposals from architects to submit plans for a new city hall.

By far, the most accomplished competitors were John McComb Jr. and Joseph Francois Mangin, a French architect. The two submitted a plan that won the design competition and received the competition prize of $350. To his credit, even before joining the new city hall project, Mangin was appointed a city surveyor and chief engineer of the fortifications in New York City as early as 1790. He also published a map of the city in 1803, and in 1815, he was the architect of the landmark St. Patrick's Old Cathedral at 260–164 Mulberry Street.

Although there is no record of Mangin's membership in the GSMT, Mangin and McComb Jr. worked together to oversee the construction of the new city hall building and design the architectural detailing. Construction, however, was delayed because the city council objected that the design was too extravagant and requested that the plans be reduced to save money. In response, McComb and Mangin reduced the size of the building and used brownstone at the rear of the building to lower costs. Some say that the city fathers, at that time, did not think that leaving the rear of the building unfinished mattered because they felt that New York City would not extend beyond that point. Little did they realize how wrong they were in making

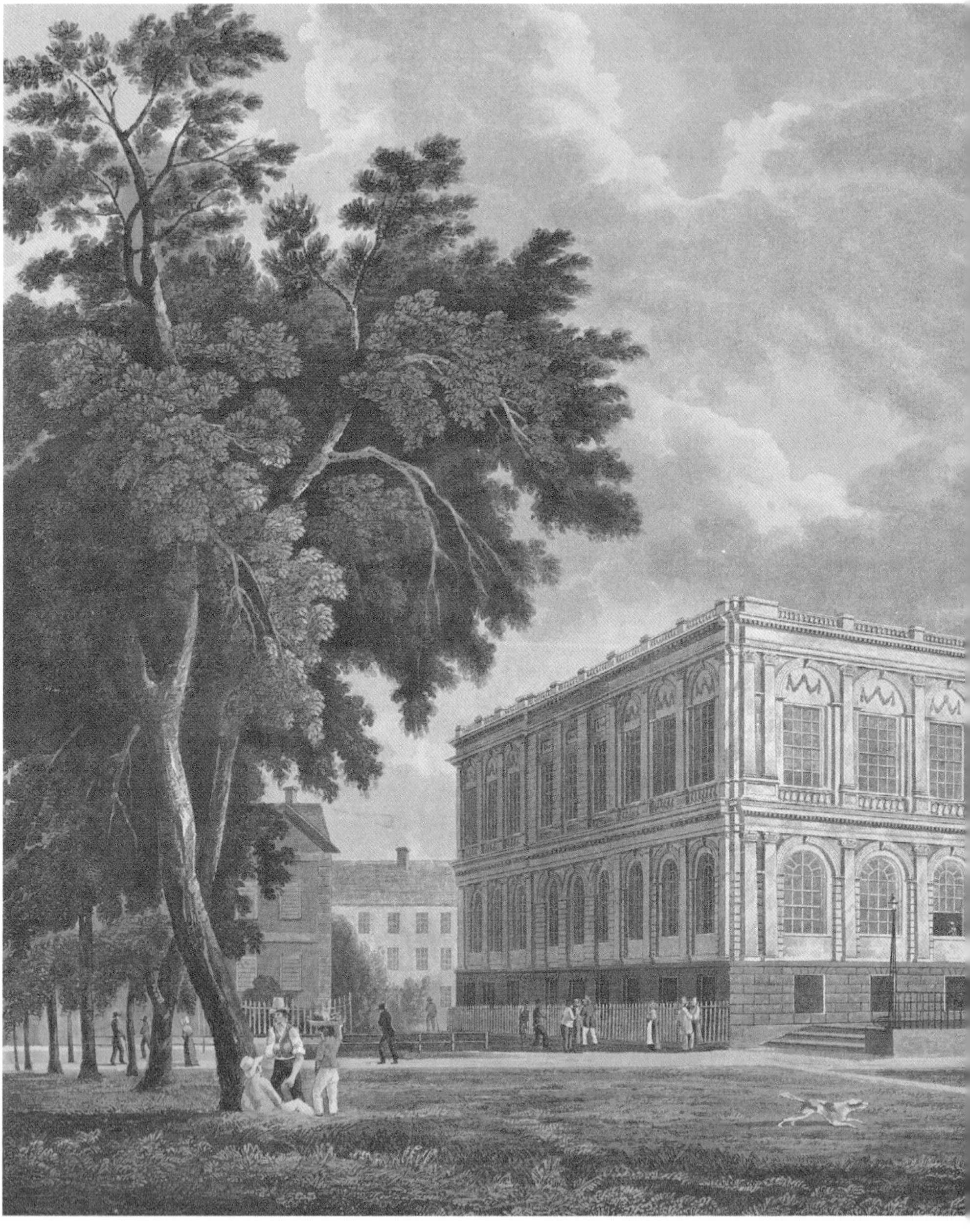

New York's city hall, 1826. Aquatint by W.G. Wall. *From the Collection of the City of New York. Courtesy of the Design Commission.*

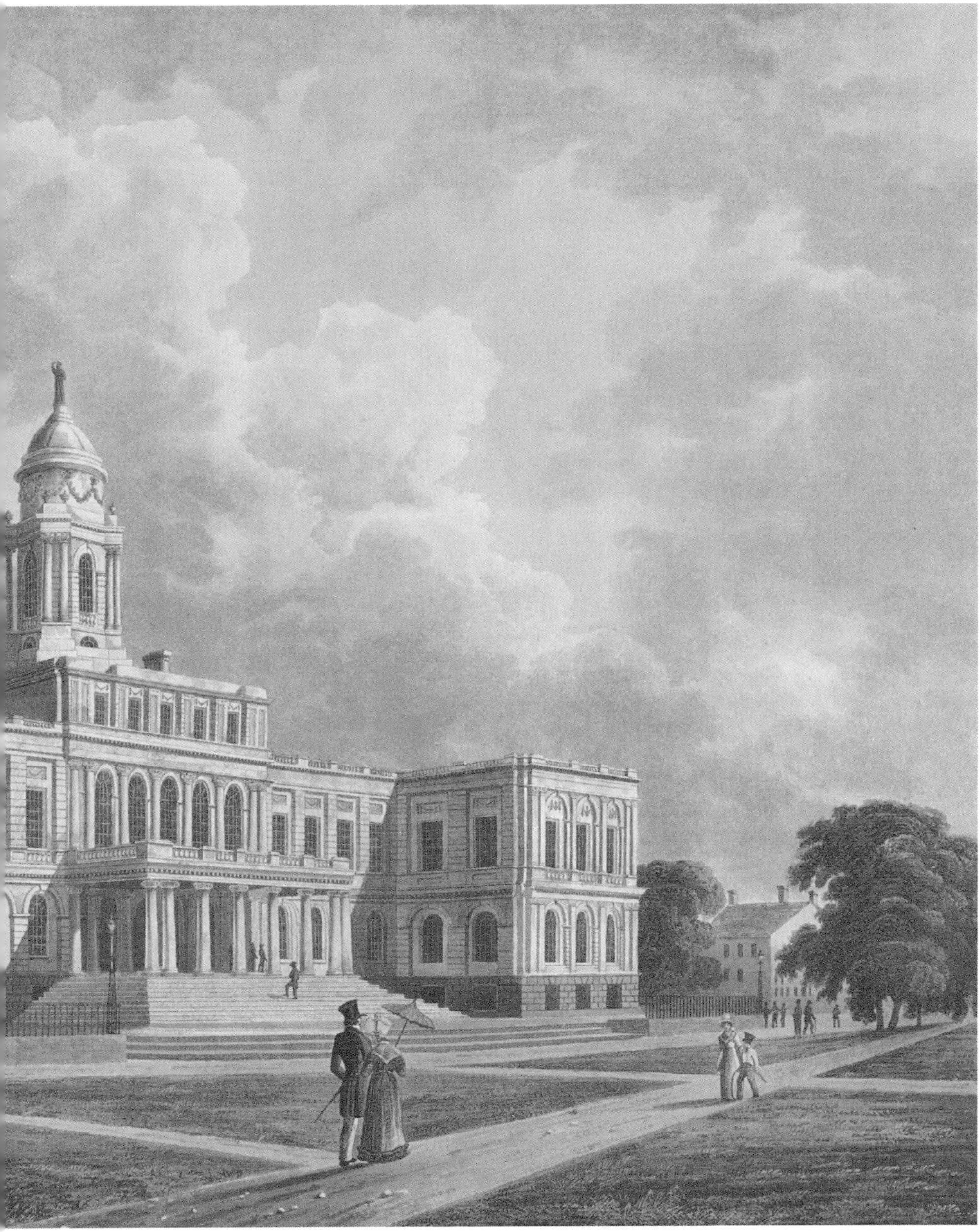

this assumption. The population was increasing, new housing and public buildings were being built, and by the time the new city hall completed, it would turn the city into a lively cultural place that was beginning to be recognized as the first city of the nation.

Although McComb and Mangin initially designed the building, without any substantial historical reference, there seems to be no clear explanation of Mangin's departure from the project. However, in 1803, the council appointed McComb the sole architect in charge of construction of city hall, and only his name is inscribed on the foundation as the architect. This generated a controversy about whether he or Mangin was the principal designer.

In addition to the architect McComb, the other key workers on this project included other prominent men. The two marble slabs identifying the men were at first placed at the southern front of city hall but later removed and erected as a mural tablet in the general lobby of the building. The marble tablet identifying the men whose hands had created one of the most beautiful scenic and historic buildings in the city is inscribed as follows:

Architect	*John McComb, Jr.*
Master Stone-Cutter	*Abraham Labagh*
Master Masons	*Anthony Steenbeck and Arthur Smith*
Master Carpenter	*Joseph Newton*
Clerk	*James Hopson*
Sculptor	*John Le Maire*

It is interesting that these men were all members of The General Society. Mr. Labagh, for instance, was president of The Society (1802) when this structure was planned, and McComb held the office of president (1818). Society members also included Joseph Newton (1790), Anthony Steenbeck (1792) and Arthur Smith (1804). The sculptor John Le Maire had joined The Society (1802) and James Hopson, Society member (1801), the clerk on this project, was The Society's secretary from 1805 to 1826.

The architectural style of city hall, constructed from 1810 to 1812, combines French Renaissance on the exterior and American-Georgian in the interior design; and the entrance of the building, reached by a long flight of steps, has figured prominently in civic events for over two hundred years. The building—located at 260 Broadway and Park Row in lower Manhattan, New York City—is the oldest city hall in the United States and still houses its original governmental functions, such as the office of the mayor of New York City and the chambers of the New York City Council.

On another note of historical significance, demolition was on the agenda for old Federal Hall. It was demolished in 1812 to make way for a new building erected in 1842, which served as the United States Customs House. In reverence to its role as the site of the first Congress, the building is now designated as the Federal Hall National Memorial, with John Quincy Adam Ward's statue of George Washington facing Broad Street and the New York Stock Exchange at 26 Wall Street.

Robert Fulton

Robert Fulton (1765–1815), a Pennsylvania-born Irish American, may not be listed in The Society's membership archives, but his name comes up in historical references from The General Society itself with an account of his achievements. An inventor, an engineer and a painter, Fulton had studied with the famed painter Benjamin West in London but subsequently he was drawn elsewhere and switched from the study of painting to civic engineering.

Fulton is best known for his steamboat, but when the word spread about his invention, it met with opposition and jeers. One member of the assembly in New York in the winter of 1797 exclaimed:

> *Fiddlesticks! Do you want to be the laughing stock of the whole country? A steam boat! Ridiculous! You are really not serious in asking us to grant a patent for a steamboat to ply the Hudson and all other island waters of the country, are you? It's some crazy scheme to propel a boat without the use of sails, I believe. The vessel is to go regardless of wind or tide.*

Skepticism was so rampant at the time that along the riverfront and in the marketplace, people called it "Fulton's Folly." Notwithstanding, in 1803, Fulton tried out an experimental steamboat on the Seine River in Paris, France, with convincing success. Then in 1806, with the encouragement and collaboration of Chancellor Robert R. Livingston, he moved to New York to build the world's first steam-powered boat.

Fulton's connection to The General Society may best be explained by the fact that he engaged the services of a millwright, the Scotsman Society member (1810) Robert McQueen. Fulton also brought a local coppersmith aboard on the project, and from England, he imported a powerful engine by Bolton & Watt, the same company in which John Hewitt, the cabinetmaker, first worked as a mechanic in America. Readers may wish to note the fact that John Hewitt was Abram Steven Hewitt's father. Both John and Abram Hewitt's association with The Society is discussed in subsequent chapters.

At last, the great day of launching arrived. Fulton's *North River Steamboat* cast off from a dock at Cortlandt Street, the namesake of the Van Cortlandt family, who had amassed a fortune as brewers. To an awaiting crowd of supporters and naysayers, the fledgling launching of the steamboat came after false starts. It docked at Clermont, New York, on its maiden voyage, thus acquiring the name the *Clermont*, and was greeted by the cheers of victory from viewers on the riverbank. Replicas and memorabilia from the legendary steamboat are housed at the Clermont State Historic Site in Germantown, New York, the location of the elegant estate of the distinguished Livingston family.

The first commercially successful passenger steamboat in the world began service on August 17, 1807, and continued in service for nearly thirty years. Fulton went on to design a steam vessel for the United States Navy as well as other ships and ferryboats.

Duncan Phyfe

The new nation and its growing population opened up opportunities for skilled artisans to supply household furnishings, and the making of fine furniture was a highly respected craft. Master cabinetmakers and their apprentices, joiners and carvers provided whatever furniture was needed for the home—tables, bedsteads, highboys, lowboys, bookcases, secretaries and chairs. English pattern books showing designs in Empire, Sheraton, Regency, Federal and French Classical styles were sources of easy reference.

Duncan Phyfe, however, one of nineteenth-century America's leading furniture makers, carved a distinctive style of his own, a unique blend of the English Neoclassical and Regency styles. Like other craftsmen, he interpreted the fashionable trends of the time and influenced other cabinetmakers, including John Hewitt (Abram Steven Hewitt's father).

Duncan Phyfe (1770–1854) was born Duncan Fife in Loch Fannich, Scotland. Duncan's family immigrated to America in 1784 to Albany, New York, where "he had labored among wood shavings as an apprentice to a cabinetmaker."

The training of an apprentice was a major step in the process of becoming a master cabinetmaker, and that required highly technical skills. Duncan

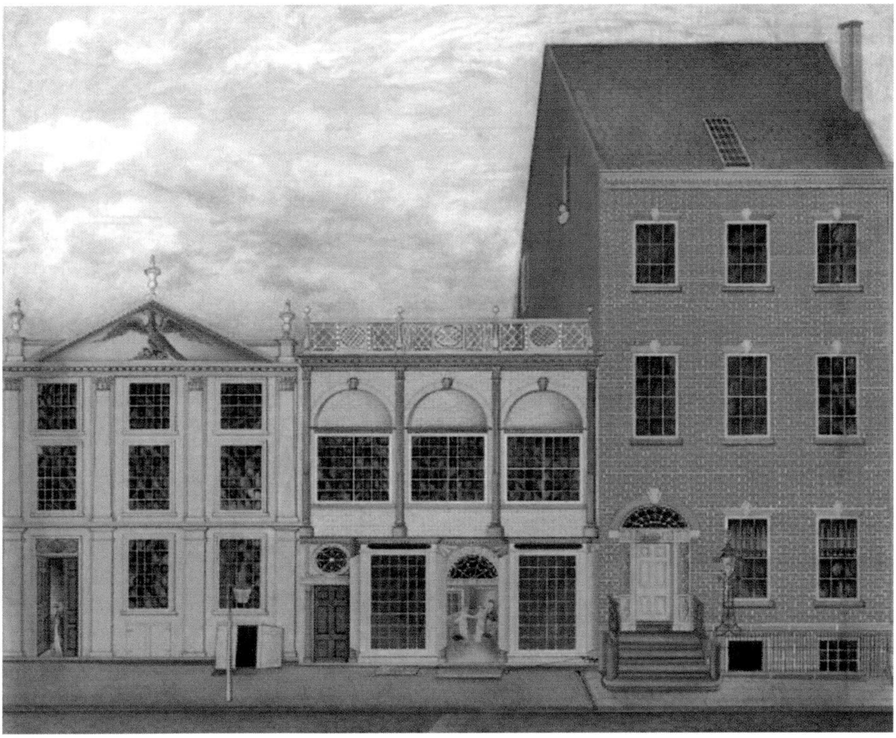

Duncan Phyfe's shop and warehouse at 168–172 Fulton Street, New York City. The image was formerly attributed to John Rubens Smith, circa 1816. *Courtesy of the Metropolitan Museum of Art, www.metmuseum.org.*

was on the move toward establishing himself, and the earliest mention of his presence in New York City is in 1791. In 1792, the year of the GSMT's charter, he is listed as a member of The General Society.

It appears that Phyfe was proposed as a candidate for membership by two men for whom he may have worked as an apprentice and either one of whom could have trained him—GSMT member Morris Nichols, listed as a member (1792) and seconded by Seabury Champlin, who is also listed as a member of The Society (1791). Champlin served as president of The Society (1797) and was an original eighteenth-century charter member of the GSMT.

Phyfe moved from his first shop on Broad Street to Partition Street, later renamed Fulton Street in honor of Robert Fulton, where he set up his shop and warehouse at 168–172 Fulton Street. About this time, he changed his name from "Fife" to "Phyfe," which suggested a more elegant, classical derivation. The new location in New York's great port of commerce facilitated the shipment of his furniture to the southern states in addition to destinations in Latin America. His greatest contribution was his personal interpretation of style, a blend of English and French furniture, and he may have referenced the illustrated pattern books such as Thomas Sheraton's *The Cabinet-Maker and Upholsterer's Drawing-Book* (1793).

Duncan Phyfe's career spanned six decades, during which the cabinetmaker organized the finest large-scale operation by merging various specialties associated with furniture production under one roof. His workshop crafted sophisticated wares that were sought after in the early nineteenth century for the dining rooms and parlors of the city's elegant dwellings. The November 11, 1917 *New York Evening Post* reported:

> *Now I will venture to assert, with entire confidence, that the best, the very best household furniture…whether taking into consideration the materials, the workmanship, or the taste and elegance of the design, has been made here. And whoever wishes to be satisfied of the degree of perfection to which our mechanics have arrived, may gratify their curiosity by calling at any time, at Mr. Phyfe's cabinet ware-house in Fulton-street, and looking at his articles of cabinet work.*

Phyfe chiefly used mahogany and introduced classical elements into his furniture, including his favorite motif, the lyre and harp. His chairs were often arched in the manner of the Greek Klismos chair and ornamentation related to the decorative arts of classical antiquities. The rapid economic revival after the War of 1812 brought about a period of peace and prosperity in New York and, with it, an increasing demand for fine craftsmanship. Phyfe's work took on an air of exclusivity, attracting even the Astor family, which started the Duncan Phyfe vogue.

America's most notable cabinetmaker had a hand in greatly influencing the refined taste of early New Yorkers who coveted owning his furniture. One lady, Sarah Elliott Huger, wrote to a friend in 1816, "I waited in company with a dozen others, at least an hour in his cold shop, and after all, was obliged to return home, without seeing the great man."

Phyfe's furniture today is highly coveted by collectors and museums. It is fitting, therefore, to remember that Duncan Phyfe, America's foremost cabinetmaker, is also revered in historic places. For instance, an elegant, green-and-white-striped-satin Phyfe sofa is part of the furnishings in the Green Room at the White House in Washington, D.C.

Popular culture also refers to Phyfe's elegant style. In one of the passages of the 1934 novel *Tender is the Night*, the author F. Scott Fitzgerald wrote "She wept all over a set that cost a fortune, in a Duncan Phyfe dining-room."

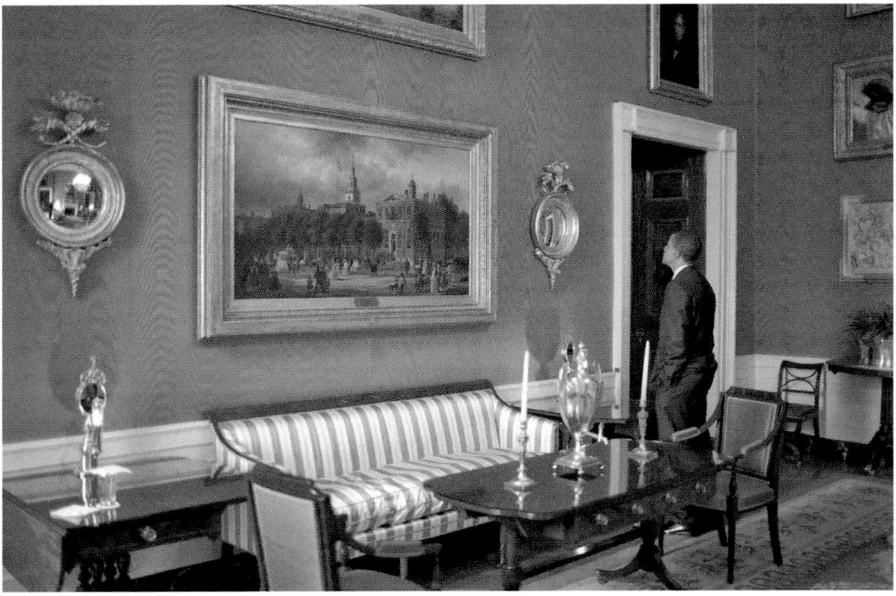

A Duncan Phyfe sofa in green and ecru striped silk in the Green Room of the White House. President Barack Obama is standing nearby, 2009. *Photograph by Pete Sousa. Official White House Photograph.*

Many items celebrating New York's history are housed in The Society's museum. Included among the trove is the official medal, engraved by C.C. Wright, celebrating the completion of the Erie Canal. It is housed in a Duncan Phyfe–designed wood box and was given by the State of New York to The General Society at the opening ceremony of the Erie Canal in 1825.

PHYFE AND HEWITT CABINETMAKERS

Phyfe's popularity, craftsmanship and signature style had a marked influence on other noteworthy artisans. John Hewitt, for one, a noted cabinetmaker in this era, was often referred to as a rival of Duncan Phyfe. Hewitt came from a long line of cabinetmakers and had immigrated to America from Staffordshire, United Kingdom, in 1796 and is listed as a mechanic and cabinetmaker. The thriving environ of Old New York was the ideal setting for commerce, and in 1805, Hewitt opened a small shop at 191 Water Street, between Beekman and Burling Slip, which had been named for the merchants, brothers William Slocum Burling and Samuel Burling, who had a glass and cabinet warehouse there.

Hewitt had been trained by his cabinetmaker father in England and brought to his craft well-documented skills in the furniture he produced for clients in North and South America. He used the Burling slip for shipping, and in the early years of his business, he thrived there as a successful furniture entrepreneur. During the halcyon days of Hewitt's flourishing cabinetmaking business, he exported extensively to the south, and the principal outlets for that market were Savannah and St. Marys, Georgia. During this time of entrepreneurial success, John Hewitt became a member of The Society in 1810.

JOHN HEWITT

Early nineteenth-century craftsmen working in the style of Phyfe, like John Hewitt, provided high quality wares, evidently with their own stamp of interpretation. It is curious to note that examples of Hewitt's and Duncan Phyfe's furniture were showcased in a Metropolitan Museum of Art exhibition, "Duncan Phyfe Master Cabinetmaker in New York," which I visited in 2011, and I made particular note of the Hewitt furniture that had marked similarities to Phyfe's work. The furniture exhibited attests to Hewitt's ancestral skill to closely model pieces in the popular style of the day. For instance, in 1812, his records indicate a customer's request for "a French sideboard like Phyfe's."

Hewitt would have continued in the enterprise of cabinetmaker as he had achieved a certain degree of notoriety and was an accomplished artisan in his own right with a substantial following in New York as well as the southern states. Unfortunately for Hewitt, the yellow fever epidemic curtailed shipments to the south and caused business to decline. The sales he garnered resulted in meager compensation. Even Hewitt's famous competitor Duncan Phyfe sold furniture at a loss, far below its worth. Adding to this misfortune, a fire destroyed John Hewitt's home and workshop on Water Street, and he was left destitute. However, an uncle came to the rescue and brought the family to Haverstraw, New York. That is where John Hewitt's son, Abram Stevens Hewitt, was born. Abram would become a mayor of New York City (1887–88) and had close ties to Peter Cooper, The Cooper Union founder. Both men became members of The General Society.

Mayors of New York

Stephen Allen

There were other significant members of The Society with illustrious backgrounds who served as mayor of New York City, including Society member (1792) Stephen Allen. He began his career as an apprentice for a sail maker during the American Revolution and picked up the art of cutting sail. It was the great era of the shipping trade, and opportunity abounded for those who would work hard. Allen was just such a character and went into business for himself and amassed a sizeable remuneration for his services. Later, he formed a partnership with Society member (1800) Joseph Lathrop.

Allen became a director of The Society's Mechanics Bank and was elected mayor for the three terms (1821–24). A key to his success was the maxim "Not to be in debt and if possible to avoid it. An artisan should work diligently and honestly, remaining within his means."

James Harper

James Harper and his brother established the J. and J. Harper and Brothers publishing house in 1817. James became a member of The Society (1831) and was elected mayor in 1844 as the nominee of the American Republican party with Whig and Democratic support.

Jacob Aaron Westervelt

Jacob Aaron Westervelt was a master of his trade and, for fifty years, owned a shipbuilding firm on the East River that produced about two hundred seagoing vessels that sailed out of the great port of New York. All along the waterfront, from South Ferry to Dry Dock at the foot of Twelfth Street, were splendid deep-sea-going ships in every stage of construction.

Westervelt joined The Society (1831) and was elected president in 1845. Later, he gave The Society all the models and pictures of the celebrated vessels he had built. Westervelt served as mayor from 1852 to 1854, during which time he introduced reforms and reorganized the police department.

Although they were not members of The Society, I would be remiss if I did not mention that some of the world's greatest shipbuilders in New York included Christian Bergh; George Steers, destined to become famous as the designer of the yacht *America*, built for John R. Stevens of Castle Point, Hoboken, New Jersey; and Bell & Brown, who built North River steamboats.

Dozens of other notable examples, amid hundreds of members over the years, made significant contributions to The General Society and, through their industry, contributed to New York's history and industrialization.

5

MEMBERS INSPIRED BY THE GENERAL SOCIETY

The influence of this Society was very great when it moved one of your members to found an institution like The Cooper Union.
—Abram Stevens Hewitt

I still feel somewhat in debt to the world.
—Peter Cooper

The success of The General Society had a great influence on many of the pioneers of American education and in the establishment of institutions that bear their name. Two such men were Peter Cooper and Abram Stevens Hewitt. These like-minded, self-made men had a strong bond that linked them in business as well as family matters. They were both members of The General Society and were inspired by its philanthropic advances and educational programs.

The ancestral history of these two men goes back to Peter Cooper's grandfather John Campbell, who was one of the founding fathers of The General Society and its second chairman in 1787 and 1788. Another family member, John Hewitt, the cabinetmaker, was Abram Stevens Hewitt's father. John is listed as a member of The Society in 1810. Rounding out the family's legacy, Peter Cooper joined The Society in 1837, and Abram Stevens Hewitt joined in 1888. As members of The General Society, Cooper and Hewitt are best remembered for establishing the tuition-free educational institute, The Cooper Union.

ABRAM STEVENS HEWITT

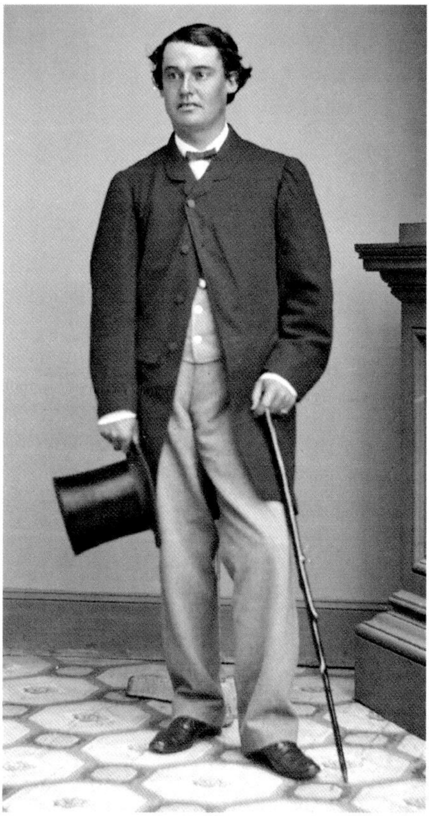

Society member (1888) Abram Stevens Hewitt, mayor of New York from 1887 to 1888. *Courtesy of the Library of Congress.*

Abram Stevens Hewitt was active in philanthropic affairs throughout his highly productive life. He built his reputation as one of the foremost ironmasters in America and received the Bessemer Gold Medal for contributions that advanced the iron trade, which secured for him an important place in the Industrial Revolution. He was an accomplished businessman and was elected to the United States Congress in 1874 and continued in that capacity until he defeated Theodore Roosevelt in the campaign for mayor of New York, serving but one year (1887–88). At the height of his political and business success in 1888, Hewitt applied for membership in The General Society.

The connection between Abram Stevens Hewitt and Peter Cooper is part romantic story and part business partnership wherein Hewitt became a close confidant to Peter Cooper. To backtrack but a minute: Hewitt had been awarded two scholarships to Kings College (now Columbia University). During his college days, he funded his additional expenses by offering a tutoring service. As the good fortune of fate would have it, one of his clients was Peter Cooper's son, Edward Cooper, who had fallen behind in his studies due to illness. During that time, the two young college men became friends and embarked on a European study tour. On returning to New York, their ship sank and with little hope of survival in a broken-down boat, they almost gave up hope of rescue. Miraculously, they were saved from the icy waters.

After that near-death episode, Abram Hewitt and Edward Cooper forged a strong friendship and went into the iron business together, and formed

Cooper Hewitt & Co. By now, Peter Cooper's daughter Sarah Amelia, the once gangly pigtail girl she was when Hewitt first met her, had blossomed into an attractive Victorian young woman. Hewitt was a frequent favorite in Peter Cooper's circle and became a valued family member with second-son status when he married Sarah Amelia in 1855. They all lived together in a sprawling house at 9 Lexington Avenue near Gramercy Park.

Hewitt considered the most successful feature of his career the organization of The Cooper Union, which he aided Peter Cooper in founding in 1857. The Cooper Union may have been Peter Cooper's great vision, but it was Abram Hewitt who took a leading part in establishing the "free" college. He supervised the construction of the building, was its financial manager and served as president of The Cooper Union for over forty years.

Abram Hewitt's cultural influence on his family is an interesting story, as is the fact that Peter Cooper, who always wanted a museum in The Cooper Union, was the grandfather of Hewitt's daughters, Amy Hewitt Green, Eleanor Garnier and Sarah Cooper Hewitt. The sisters, adhering to their grandfather's vision and their own interest in the decorative arts, created the Cooper Union Museum for the Arts of Decoration in 1897 on the fourth floor of The Cooper Union. The Hewitt sisters—Amy, Eleanor and Sarah—were three remarkable women for their time, but after Amy married, Eleanor and Sarah, the principal stewards of the museum, spent their life and their fortunes on the museum's collections.

Due to the expansiveness of the collections, the museum eventually fell under the ownership of the Smithsonian, and in 1994 the museum's name was changed to the Smithsonian, Cooper-Hewitt National Design Museum in New York City and is housed in the elegant and commodious Carnegie Mansion in New York. Another name change took place in 2014 when the museum was renamed the Cooper Hewitt, Smithsonian Design Museum.

Peter Cooper

The industrialist, inventor, manufacturer and philanthropist Peter Cooper lacked a formal education, which he regretted his entire life, but he was gifted with keen intelligence, the skill of invention and an entrepreneurial drive. As a young man, Cooper entered as an apprentice to New York coach maker John Woodward of the firm Burtis and Woodward, which was a great business at that time. He remained with the firm for four years until he had thoroughly learned the trade and then moved on, working at or buying other businesses. Among Cooper's many inventions was a machine for shearing cloth, which proved very profitable during the War of 1812 when textiles were curtailed from Europe. The foundation of his fortune, however, was gained by the manufacture of glue and his ironworks, occupations he held at the time of his initiation into The General Society in 1837. Despite the fact that he was one of the city's millionaires, Cooper is listed in The General Society records as a "glue maker," a tradesman term with all modesty that he preferred.

Cooper's Canton Iron Works in Baltimore built the first steam engine. He was a pioneer in communication technology and helped to finance the first transatlantic telegraph cable. A man of remarkable health and fortitude, in his eighties, he mounted a campaign for the office of president of the United States, running unsuccessfully as the candidate on the Independent Greenback Party in 1876.

Throughout his life, Cooper was an ardent supporter of free education for the working classes and did not want any other person to suffer the disappointment he had for lack of a formal education. Cooper was influenced

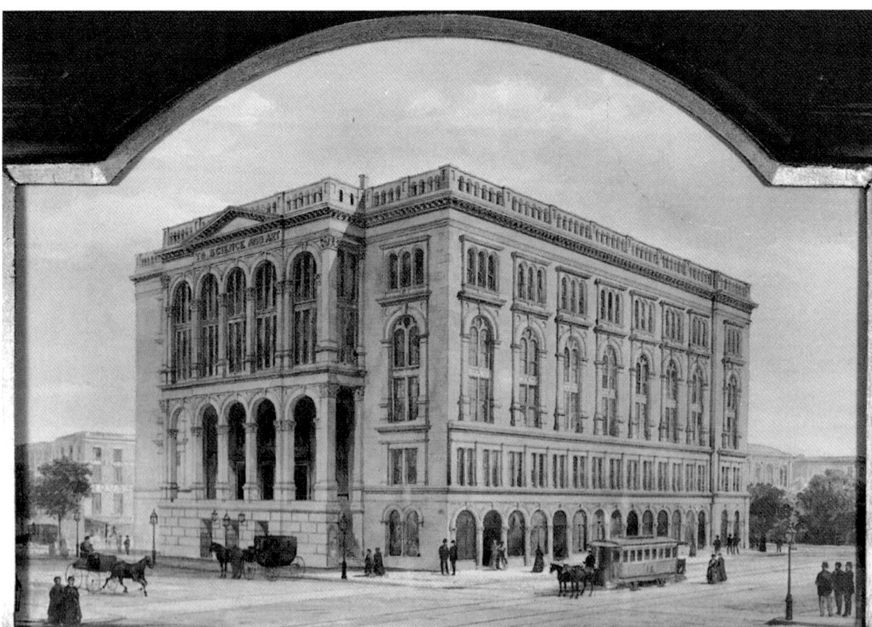

Cooper Union for the Advancement of Science and Art located at Third and Fourth Avenues, Seventh Street, New York. *Courtesy of the Library of Congress Historical American Buildings Survey.*

by the educational advantages that The Society gave to the apprentices of New York and was determined to carry out the objectives of The Society in a school of greater magnitude.

To this end, he founded The Cooper Union for the Advancement of Science and Art. The cornerstone was laid in 1857. In 1859, it was charted by the state legislature and renamed The Cooper Union. Cooper's aim was to provide free education for working-class male and female students in the fields of engineering, architecture and domestic science. Cooper's inspiration to offer college level studies at The Cooper Union may well have had its kernel of inspiration from The General Society's Mechanics Institute's educational program itself. Attesting to this fact, the following excerpt from a letter to The General Society from Cooper's son-in-law Abram Stevens Hewitt suggests such a premise:

> *The History of the institution (Cooper Union) is one of great beneficence, and at no time has its work been more valuable to the community than the present. I am quite sure that the late Peter Cooper derived a portion of his ideas from his connection with The Society of Mechanics and Tradesmen, that he based the instructions departments of The Cooper Union upon the experience which he had gathered in your schools.*

Throughout Cooper's long life, he called himself "a mechanic of New York," and with his frugal posturing—as he rode about New York in an old-fashioned one-horse chaise drawn by a steady mare—he appeared more like a wealthy farmer than the millionaire that he had become. Cooper had his hand in many of the country's scientific and industrial undertakings, and it was his money that made possible the years of experimentation in laying the Atlantic Cable. As Penny Speckter said, "It was the 19th Century equivalent to putting a man on the moon." A mounted slice of the Atlantic Cable (1858) is on display at The General Society's headquarters. Modest to the end, before he died at the age of ninety-two, one of Peter Cooper's last comments was, "I still feel somewhat in debt to the world."

Pillars of Society

Education was of paramount interest to other members of The Society. Many of these men had immigrated to America poor and unskilled, with nothing more than their brilliant minds, their faculty for innovation and a determination to succeed. These self-made men became pillars of New York society. The legion of success stories would fill volumes. For the sake of brevity, following are other GSMT members, the Brothers, who made their indelible mark on history.

William Colgate

William Colgate arrived in New York as a boy, with only the knowledge of soap and candle making. Forlorn and looking for work, he wandered the streets of the city and fortuitously met a canal-boat captain who gave him the following advice:

> *Well, young man, let me pray with you and give you a little advice. Be a good man, make an honest soap; give full pound and I am certain you will be a prosperous and rich man.*

At first, Colgate served as an apprentice to a soap-boiler. He learned the trade, and through shrewd observation of the firm's operations, he established a starch, soap and candle business, in 1806, known as Colgate and Company on Dutch Street, the namesake of the city's founders. His company later became known as Colgate-Palmolive, and he remained faithful to the advice given to him by the captain and led a spiritual life and fervently believed in tithing in gratitude for his success. Colgate became a member of The Society in 1831 and was a leading philanthropist and helped to found Colgate University in 1890.

Charles Pratt

Born in 1830 in the factory town of Watertown, Massachusetts, Charles Pratt rose from his early training as a machinist to ownership of one of the country's largest refineries. When he came to New York City in 1851, he became a clerk at an oil, paint and glass shop on Fulton Street. By 1867, he established Pratt Manufacturing Company, producers of Astral Oil in the nearby Williamsburg section of Brooklyn. Pratt became a pioneer in the United States petroleum industry and became a very rich man when Astral Oil was later acquired by John D. Rockefeller's Standard Oil in 1874.

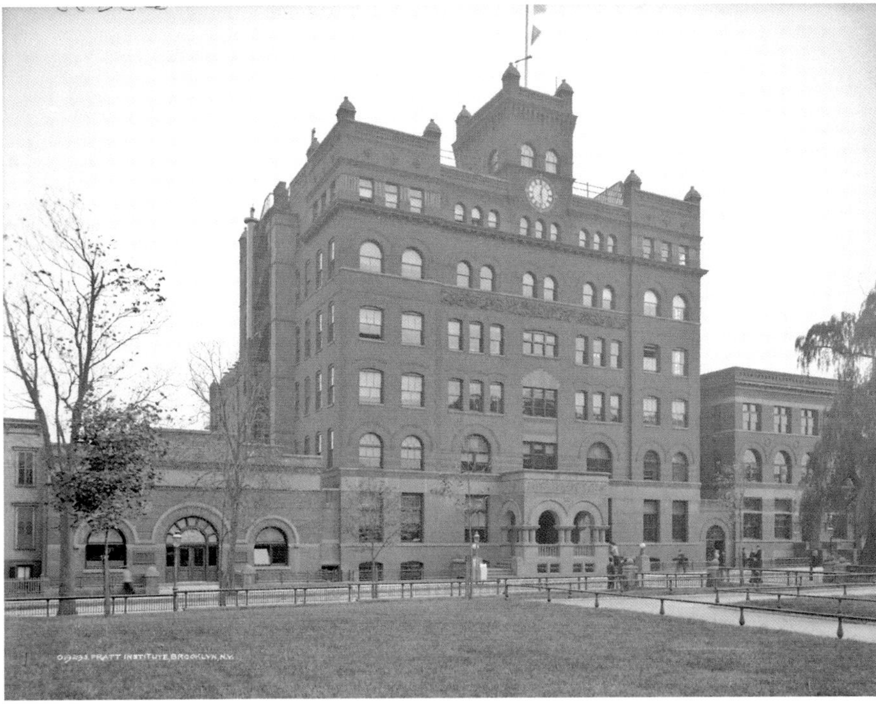

Pratt Institute in Brooklyn, New York. This is the first building (constructed circa 1892) to house the school, when twelve students arrived for the opening day, 1899. *Courtesy of the Library of Congress.*

An advocate of education, Pratt is credited with recognizing the growing need for trained industrial workers in a changing economy, and in 1886, he founded and endowed the Pratt Institute. It opened its doors in Clinton Hill, Brooklyn, New York, in 1887, the same year Charles Pratt became a member of the GSMT.

On the opening day, October 17, twelve students showed up and paid four dollars per class per term. The pioneering founder told the students, "Be true to your work and your work will be true to you." Originally a technical institute with training in manual and mechanical arts, Pratt has become a renowned art, design and architectural college. Throughout his lifetime, Pratt donated large sums of money to several institutions and felt a strong obligation to give to others, and tithing was a constant discipline. Pratt died in 1888, and in the same year, a tribute appeared in *Scientific American*: "The Institute is a monument to the philanthropy and wisdom of its founder, an ornament to the City in which it is located, and a credit to the country at large."

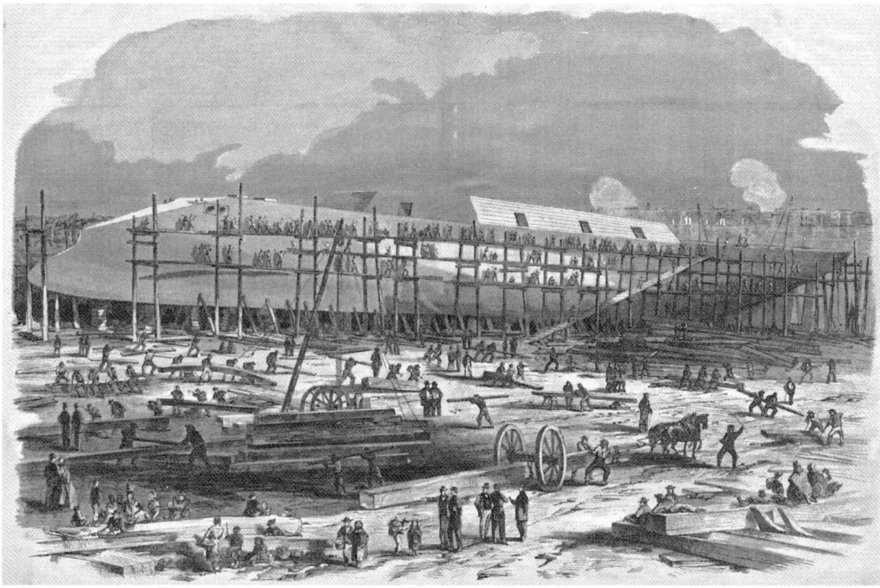

The USS *Dunderberg* under construction from 1862 to 1863. "The greatest Man of War in the world in William H. Webb's yard," as described in *Harper's Weekly*, November 14, 1863. *Courtesy of the Webb Institute.*

William H. Webb

William H. Webb, industrialist and philanthropist, established his career as an internationally prominent naval architect and was noted for designs of clipper ships and the giant ironclad USS *Dunderberg*, completed shortly after the Civil War. He was described in one contemporary tribute in the *New York Herald* as "the very first naval architect in the country."

After a successful career, he shuttered his shipbuilding facility, and with an interest in naval education, he felt that ship designers of the future needed to be skilled in structural and mechanical engineering and other scientific disciplines. To this end, he founded the tuition-free Webb Academy and Home for Shipbuilders in 1889, which today is known as the Webb Institute, located in Glen Cove, Long Island. Webb provided an endowment to allow the institute to be self-supporting in perpetuity. Students who are U.S. citizens are not charged a tuition.

In 1918, Webb announced his intention to erect a commemorative tablet on the building at Montgomery and Water Streets. The tablet recorded the fact that on that site, in the years 1824 and 1825, his father, Isaac Webb, built two ships of about six hundred tons each, the *Superior* and the *Splendid*, for the China Trade.

Isaac Webb (1823) and his son, William H. Webb (1841), were honored members of The Society.

Warren A. Conover

Conover was active in committee work of The Society and for years was The General Society's representative on the board of William H. Webb's Academy and Home for Shipbuilders. He was initiated into The General Society in 1883 and served as its president in 1895. A prominent builder in New York, Conover was also a member of the New York City Board of Examiners of the Building Department.

Richard and Cornelius C. Poillon

The eminent shipbuilders Richard and Cornelius Poillon are of particular note here as they too were members of the board of directors of the Webb Academy for Ship Builders. The brothers were shipwrights, employed in the construction and maintenance of ships. They were owners of C&R Poillon, a firm that specialized in wooden vessels, including racing yachts, Civil War gunboats, ferries, schooners, sloops and steamers. Richard was a member of the Library Committee of The Society in 1878 and 1879.

The company began as a shipyard and sawmill located in Brooklyn at the foot of Bridge Street on the East River, where it eventually employed three hundred men and also owned wharfs and a yard in Gowanus, South Brooklyn. Eventually, it held a monopoly on repair work in New York harbor. However, shipyard work in New York began to decline as overland transportation by rail increased and bridges and tunnels replaced ferries and ships.

James O. Poillion was the son of Richard Poillon and became associated with the firm in 1870. After the deaths of his father and uncle, James remained at the firm until the firm closed in 1904. James was initiated into The General Society in 1892. He served as resident manager (president) of the Webb Academy, the charter of the academy requiring that a member of The General Society be among its directors.

Nathaniel Currier

Nathaniel Currier was the elder partner in the firm Currier and Ives, lithographers, formed in New York in 1857. He was originally a job printer, producing fliers, tickets and certificates for clients. In 1840, Currier began to market decorative images for the general public, and among the firm's famous lithographs is Cornelius H. DeLamater's ironclad ship the *Monitor* in combat with the *Merrimac*. He joined The General Society of Mechanics and Tradesmen in 1847.

James Merritt Ives joined the firm as Nathaniel Currier's bookkeeper in 1852. The firm became known for its reasonably priced decorative images and developed a wide market in America and Europe. It eventually became the most successful lithography firm in the United States.

THE LARGESSE OF DONORS

It is interesting to note that while some prominent men may not have been members of The General Society, they nonetheless deserve recognition for their belief in the future of The General Society through their generous donations.

Amos F. Eno

Amos F. Eno may not have been a member of The Society, but he was one of its foremost benefactors. He was a prominent hotel builder and property owner whose bequest of nearly $2 million to The General Society was one of the largest donations in its history.

Amos F. Eno, the son of Amos R. Eno, had opened the Fifth Avenue Hotel at the corner of Fifth Avenue and Twenty-third Street in 1859. It was hailed as the most spectacular building of its kind erected in the city. Its wonders included private bathrooms for every room and the first passenger elevator in a hotel, a steam operated affair known as the "vertical railroad."

Amos F. Eno's interest in The General Society came about as he observed some of the workers. Among the trusted employees was William Whitenack, an Old Dutch carpenter whom Eno admired for his skill and thrift. William became a member of The Society in 1910, and it was William who suggested that Amos Eno help the Mechanics Institute.

Eno was obviously impressed with the organization's raison d'être, and in 1915, he presented The General Society with two buildings valued at $100,000. When he died later that year at the age of seventy-eight, it was found that he had also left The Society another $1,800,000.

However, as with many large legacies, the will was contested. Questions of Eno's soundness of mind were raised. The case dragged on for several years, and not until 1924 was the final legacy paid to The Society, along with accumulated interest. It was finally recognized that the legacy to The Society was in accordance with Amos Eno's long expressed sentiments of desiring to help people who work with their hands. Upon the final settlement of the provisions of the Eno will and payment to The Society of this great endowment, work was immediately begun to increase classroom space and equipment, provide fire exits and replace the obsolete wooden shelving in the Library with steel bookcases.

Throughout the years, The General Society has been the recipient of generous financial and historical donations, as well as funds for improvements to the GSMT building from members and benefactors, a policy that continues to the present time. Garrison Keillor, the celebrated storyteller, was the headliner at a benefit. The author Thomas Wolfe also helped to raise money to restore the GSMT's library skylight, which had been black since World War II. Brooke Astor donated money for the lighting.

In The Society's history, a host of other members gave generous donations or provided financial security for The Society. Other donations of great significance include one given by Robert Boyd V, the great-grandson of the first chairman of The Society, Robert Boyd (1785 and 1786). The younger Boyd presented a three-foot portrait of his ancestor to The General Society, which hangs in the executive offices at the GSMT headquarters. The painting is a copy of the original oil, which is owned by the Brooklyn Museum. The GSMT painting is among the organization's historical treasures as it is the only likeness of Boyd, whose link to the founding of The Society is forever revered.

6

Andrew Carnegie and Distinguished Presidents

I would rather be the grandson of one who could teach this art than the descendent of thirty worthless Dukes. There is no heritage like that of being born poor. The leaders and teachers of the Nation come from the poor.
—Andrew Carnegie

Throughout The Society's 230-year history, by 2015, there were 216 presidents who had presided over the GSMT's historical development, and several of these individuals had served multiple terms. In the early years, men of great wealth, although they were not presidents, also took a profound interest in The Society. One such individual was Andrew Carnegie, who was initiated as a member in 1891.

A man of great gestures, Carnegie devoted most of his life to large-scale philanthropy, with emphasis on education, local libraries, peace, scientific research and music. However, let us not forget, that it was Peter Cooper's philanthropy that inspired the charitable acts and civic responsibilities of other tycoons.

Carnegie wrote "The Gospel of Wealth" and preached the obligations of the wealthy to give away funds for the common good of humanity. As a benefactor to libraries, he funded nearly three thousand around the world.

The largesse of Brother Andrew Carnegie to The General Society was substantial and given so generously to meet the needs or to expand The Society's building. In 1903, he gave $250,000 for the purpose of making alterations and expanding The General Society's building for larger

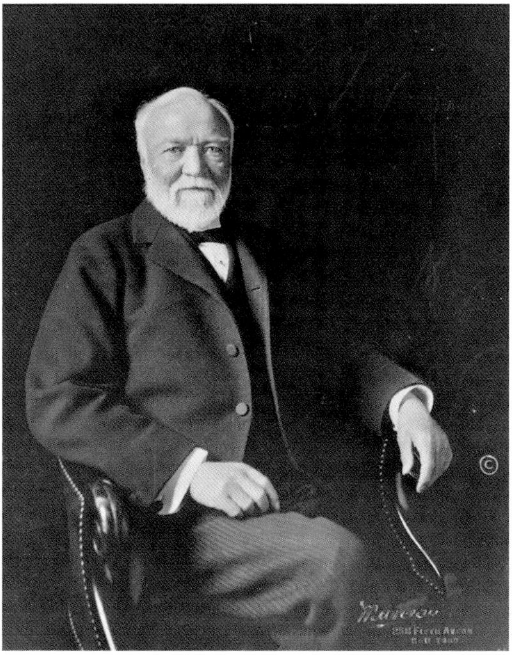

Andrew Carnegie, the GSMT's great benefactor. His gifts enabled The Society's building expansion in 1903. Portrait from Marceau, NY, 1913. *Courtesy of the Library of Congress.*

accommodations to the school. Finding that this sum was insufficient to carry the work to completion, he increased his gift the following year by $50,000. Unstoppable in his donations, in 1907, he gave a special donation of $2,000 to help make up a sum sufficient to carry on the work of the Institute. In 1908, he gave an additional $200,000 as an endowment fund, the interest to be used only for the educational purposes of The Society. Including his former gift in 1899 of $25,000 to place The General Society building "in condition for our needs at that time," this makes a total sum contributed by Brother Carnegie of $527,000. The philanthropist may also be remembered for his cultural contribution to the city, as he built New York's Carnegie Hall and many public libraries.

CARNEGIE "GLAD TO BE A POOR MAN'S SON"

Born to modest circumstances in Dunfermline, Fife, Scotland, in 1835, Carnegie immigrated to the United States with his family when he was thirteen. His is a true rags-to-riches story. Andrew's family had to borrow money in order to immigrate to Allegheny, Pennsylvania. At first, he became a bobbin boy in a Pittsburgh cotton factory. The railroads opened up new opportunity, and in 1850, he became a telegraph messenger boy. With rapid advancement, he was appointed a secretary/telegraph operator. By the 1860s, Carnegie began his foray into speculating and wisely invested in railroads, bridges and oil derricks.

In the footsteps of history, he is best known for leading the enormous expansion of the American steel industry in the late nineteenth century and building Pittsburgh's Carnegie Steel Company, which by the 1890s, became the largest and most profitable industrial enterprise in the world. In 1901, Carnegie sold his company for almost half a billion dollars to J.P. Morgan, who created the U.S. Steel Corporation.

Carnegie's interest in The General Society of Mechanics and Tradesmen was in keeping with his belief in helping those who would help themselves by providing the means for them to improve themselves and their condition through educational advancement.

"Carnegie Says He's Glad to be a Poor Man's Son," reported the *New York Times* on November 19, 1905, the day after The General Society Founder's Day banquet held at Delmonico's on November 18, 1905. In an excerpt from the newspaper, Carnegie, the guest of honor, eloquently praised The Society:

> *Ladies and Gentlemen, Brothers and Sisters...All I am here to do is to laud your society, for I have never known one with such a splendid record. The first free library was established by it and not by me. I am still hoping to live long enough to establish the last. Your Society with its high aims has been duplicated in Boston. Of your members were Peter Cooper and his great son-in-law, Abram S. Hewitt, who should rank with America's best. Although I am a member of your honorable Society, I have never been a mechanic. I worked on a boiler for several months once in a cellar, and on the recommendation of Mr. Hewitt was admitted twenty-five years ago to your membership because I had worked at the case. I am really one of you by heredity. My Grandfather was a shoemaker and a printer.*

Carnegie's speech continued:

> *I thank God that in my youth I learned how to make and mend shoes. I would rather be the grandson of one who could teach this art than the descendant of thirty worthless Dukes. There is no heritage like that of being born poor. The leaders and teachers of the nation come from the poor. Thank God that that is true, as the Bible says the poor we always have with us. They are the salt and salvation of the State, on which the true democracy must stand and subsist. It is a sweet source of satisfaction to do the little I can to further the great work of your splendid institution. I prophesy, from your useful past, that you will far exceed it in the future.*

In his autobiography, Carnegie noted:

> *The Cooper Union, the Mechanics and Tradesmen Society, indeed every institution in which I became interested, revealed many men and women devoting their time and thought, not to "miserable aims that end with self," but to high ideals which mean relief and uplift of their less fortunate brethren.*

Prominent Presidents

The Society's members who were also philanthropic over the years included some of The Society's prominent presidents. A 1915 article in the *New York Times*, entitled "What Society of Mechanics and Tradesmen Is," states:

> *Its roster of over 100 presidents includes the names of many grand New Yorkers who have been prominent in its mercantile and industrial life. Among them are two former Mayors, Gideon Lee, and Stephen Allen, the latter beginning life as a sailmaker. Others of note were Gabriel Furman, John Slidell, father of Captain Slidell of Confederate fame; Sheppard Knapp, George Ireland, the first president of the Mechanics' Bank, which was organized in 1810 through the direct interest of The Society; John McComb, architect of City Hall; Benjamin De Milt, Jacob Lorillard and Jacob A. Westervelt, a famous shipbuilder in the days when the United Sates had a regular merchant marine.*

Cornelius H. DeLamater

Industrialist Cornelius H. DeLamater initiated into The General Society in 1848 and was, for forty-one years, an active and sympathetic member of The Society, becoming second vice president (1867), vice president (1868) and elected president by unanimous vote (1869).

DeLamater was born in Rhinebeck, Dutchess County, New York, and with nothing more than a common school education, he became an errand boy for the hardware store Schuyler and Sword in New York City at the age of fourteen.

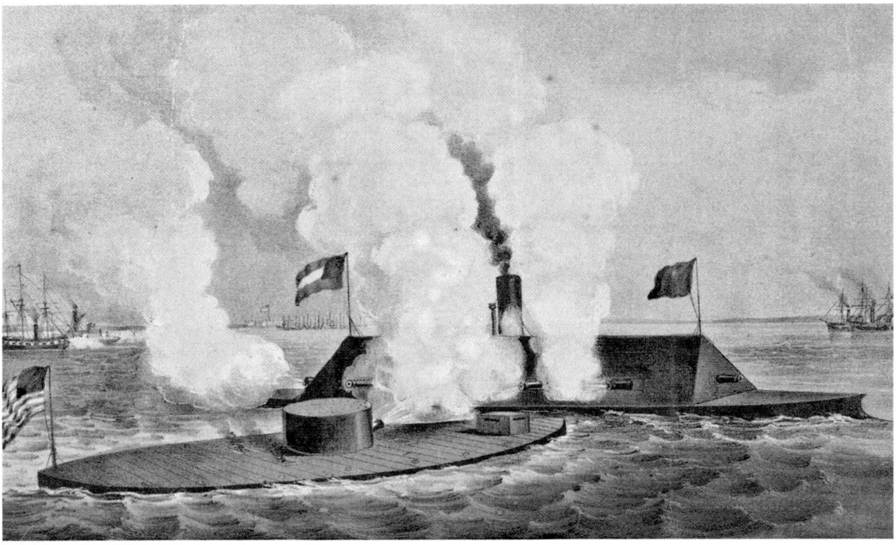

Cornelius H. DeLamater's ironclad ship the *Monitor* in combat with the *Merrimac*. DeLamater was GSMT president in 1869. *Lithograph by Currier and Ives. Courtesy of the Library of Congress.*

Then came a fortuitous opportunity. At sixteen, he worked for John Cunningham of the Phoenix Foundry where he learned the iron business. When Cunningham decided to retire, he offered the foundry to DeLamater and his cousin Peter Hogg, who was a practical mechanic. Under the new ownership, the business thrived until 1849 when DeLamater, then only twenty-eight years old, purchased the Hogg interest and built up the great DeLamater Iron Works.

Under DeLamater's direction, the business turned into a major foundry, ironworks and construction activity. He also helped shape the course of the Civil War through his support of ironclad ships for the United States. It was at Phoenix that DeLamater met the Swedish inventor and engineer Captain John Ericsson, and the two men established a business partnership that resulted in the development of some of the most revolutionary marine propulsion devices of the day.

The collaboration culminated in DeLamater's construction of the steam boilers and machinery for Ericsson's turreted ironclad USS *Monitor*. During the Civil War, the DeLamater Iron Works did a great deal of work for the government and subsequently built machinery for four additional Ericsson-designed monitors. The Society commemorated the public service of the DeLamater and Ericsson collaboration in their 1922 record book:

On Thursday, March 9th, 1922, the sixtieth anniversary of the battle between the Monitor *and the* Merrimac, *of Civil War fame, the DeLamater-Ericsson Commemoration Committee unveiled bronze tablets designed by Isadore Konti, Sculptor, on the sites of four buildings in this City, three in the borough of Manhattan and one in the Borough of Brooklyn. The Society's interest lies in the fact that the exercises of the unveiling of the bronze tablet at the second site, that of the old Phoenix Foundry, 260 West Street, was assigned to The General Society of Mechanics and Tradesmen, who cooperated with the DeLamater-Ericsson Tablet Committee, and the further fact that Mr. Cornelius H. DeLamater was President of The Society in 1869. At three o'clock on the afternoon above mentioned, a large gathering assembled at the site to participate in the ceremonies. Among those present were many distinguished guests, notably the Swedish Consul and representatives from the Navy Department of U.S. and from the Board of Aldermen of the City, also one of Greenwich Villages' oldest residents and oldest living member of The Society, Hon. John R. Voorhis, having been a member in 1860 and President in 1879.*

THE CIVIL WAR REGIMENT

Hundreds of members made contributions to the history of The General Society, and patriotism was a strong motivator. The Civil War claimed a dozen or more lesser-known GSMT members, composed almost entirely of mechanics who volunteered for the Union Army in 1861. They formed a significant part of the cadre that became the military's First New York Volunteer Engineer Regiment, the founding outfit that became the Army Corps of Engineers. On October 2, 1861, when our country found itself faced with rebellion and civil war, the following resolution was adopted by The Society:

Whereas, a regiment has been formed in our city known as the "Volunteer Engineer Regiment" which is composed exclusively of mechanics, some of whom are members of this society:

THEREFORE, RESOLVED, That this society, in appreciation of the public spirit of patriotism which originated, and the zeal and industry which perfected, the organization (now about to take the field) as a regiment of volunteers from the State of New York, from which all our chartered privileges are derived, do cause to be prepared and presented to it, under the

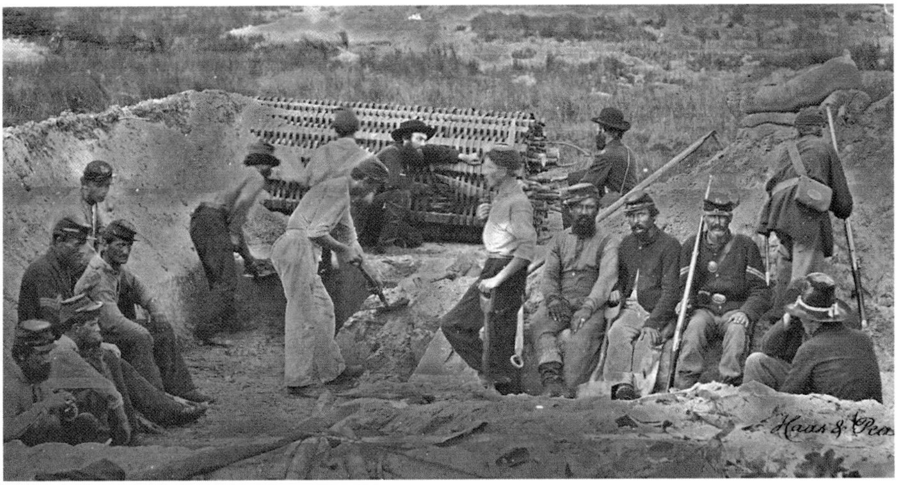

The First New York Volunteer Engineer Regiment in the Civil War in 1863 was composed exclusively of mechanics, many of whom were members of the GSMT. *Courtesy of the Library of Congress.*

> *auspices of The General Society of Mechanics and Tradesmen of the City of New York a suitable stand of colors.*

The following account was given of the presentation of the colors:

> *The flags were presented on the 23rd of January at the Seventh Regiment Armory by the president of The Society, and were received by Major Butts, in behalf of Colonel Serrell and the regiment in a brief and soldier-like response. Then the following resolution was adopted by The Society:*
>
> *RESOLVED: That thanks of The Society be and are hereby tendered to Brother Joseph Hoxie, for his most acceptable and spirited speech, delivered on the 23rd of January, on the presentation of battle-flags to the Volunteer Engineer Regiment.*

During the Civil War, as noted, The Society was ever loyal and active in the support of the government. The general condition of The Society being prosperous at this time, there was no dissenting voice among the members in regard to doing its share in the task of saving the republic. Eight thousand dollars were spent in the purchase of a government bond. The year 1865 brought the cessation of hostilities and the following correspondence was sent to Society member (1844) George W. Farnham, who served as vice president (1858) and president (1859):

George W. Farnham, Esq, Mechanics Society, N.Y.

Sir: As you were chairman of the committee who presented to my regiment a stand of colors, on behalf of The General Society of Mechanics and Tradesmen of the City of New York, when we were organized in 1861, I have taken the liberty to ask, through you, that The Society present, on behalf of the regiment, the national color of this set, and a regimental color presented by the City of New York, to the Bureau of Military Statistics at Albany, New York.

The number of engagements in which detachments of the regiment have taken part will show you that we have not been idle spectators in the great struggle. We represent almost every trade, and I believe the mechanical skill we have been able to bring together has added much to the efficiency of the army.

The flag has now become a veteran, worn out in the service, and unfit for duty in the field; and we return it, to take its place with others honored by service with our brothers of the State of New York.

The regiment color presented by you we still retain in the field, and we hope to bring it back for your disposition, after peace has been restored to the land, and we are permitted to go back to our homes and vocations of life.

I am very truly your ob't serv't and brother,
James F. Hall, Brt, Brig Genl., U.S. Vols., Col. Comdg. List N.Y. Vol. Engineers.

In a follow-up correspondence, Hall wrote, "The beautiful regimental color which you presented us is still a daily object of admiration, and as we can now add to its previous history the final battles of the Rebellion—Petersburg, Richmond and Appomattox Court House—is doubly dear to us, and I know it will be to you when we return it to your custody after our united country has told us that our work is done."

In 1868, after their return from the war and the period in which they were at the Bureau of Military Statistics in Albany, New York, the battle flags were returned to The Society. The flags are now preserved in glass cases in The General Society's Library at 20 West Forty-fourth Street.

7

Mechanics Hall and Mechanics Bank

The General Society of Mechanics and Tradesmen is justly proud not only of its origin but it history. Something of this is due to the excellent men who established The Society, or who carried through so many generations in usefulness and honor.
—GSMT annals

After years of meeting at Fraunces Tavern, the GSMT was sufficiently solvent that at a meeting held November 1, 1792, a committee was appointed to devise ways and means for erecting a suitable hall for The Society. From a report of the treasurer, John Stryker, on January 8, 1801, the assets of The Society were reported to be $7,187.41, a substantial fund due to wise investments by The Society's officers. With these funds in hand, there was talk of the possibility of acquiring a permanent home for The Society to be called Mechanics Hall. On October 7 of the same year, the committee appointed to procure a lot of land recommended the west side of Broadway, at the Corner of Chambers Street owned by a Mrs. Henshaw. However, the committee did not come to an immediate decision due to other matters concerning their benevolent operations.

Benevolent Operations

Now with sufficient funds, the fundamental purpose for which The Society was created was under consideration. The benevolent operations of The

Society—whose primary object was mutual aid, assistance in case of sickness or distress and care for widows and orphans of members who died without property—occupied considerable time during the regular meetings.

This part of The Society's work was carefully managed by what was first known as the "Board of Overseers of the Indigent" and later as the Pension Committee, whose duty was to exercise a fatherly care over The Society's beneficiaries. The committee bestowed deserving consideration on its members and widows of members who had encountered hardships. From the fund, a portion was distributed annually among the widows on the pension roll. At this time, The Society was taking care of fourteen widows and twenty-seven children and helping an occasional hard-pressed Brother with his problem.

Although The Society was permitted to loan money to members, in reference to these appropriations and apparently in a demonstration of respect to the recipient of such funds, it required that new members of The Society "on its private transactions be as silent as the grave."

On one occasion, Brother William H. Webb, creator of Webb Institute, donated an oil painting to The Society, *From Riches to Poverty*, which hangs in the GSMT's Assembly Room. Its touching beauty serves as a constant reminder of the early aid given by The Society. In his letter of donation, Webb stated:

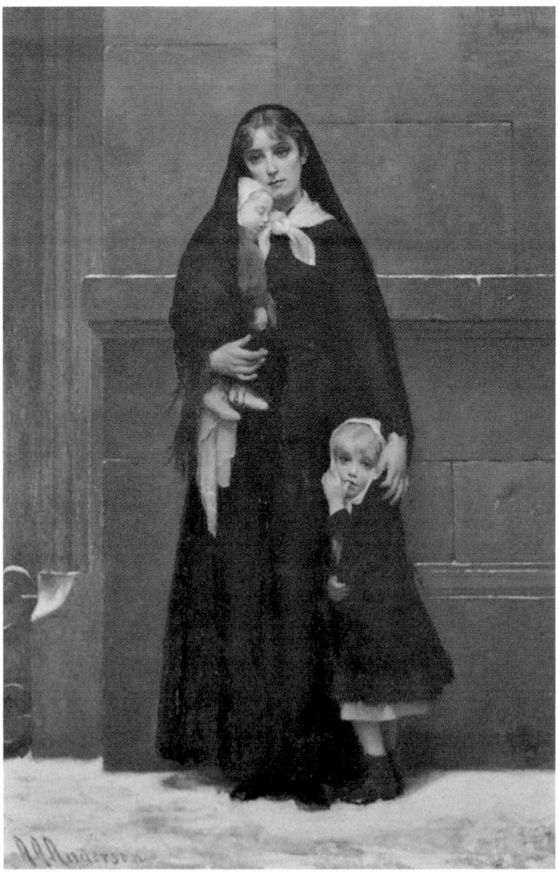

From Riches to Poverty, by Abraham Archibald Anderson, portrays a woman with a small child. It relates to The Society's founding mandate to care for its members' children, painting circa 1885. *Courtesy of Stephen Amiaga.*

I understood Mr. Anderson (the artist) to say he was prompted to execute the work by seeing something similar in the streets of Paris, in which city it was painted. The title he gives is From Riches to Poverty. *On my visit to the art gallery at New Orleans last winter, I saw this picture on its walls and was almost moved to tears. I then and there determined to posses* [sic] *the painting and present it to the Mechanics and Tradesmen Society in hope that it might forever be a reminder to its members of the charity due to the widows and orphans of their deceased members.*

FIRST MECHANICS HALL

At a session on May 5, 1802, the question of owning The Society's own hall for meetings was again under consideration. The plot of land now recommended was at the northwest corner of Broadway and Park Place (then Robinson Street), at 239 Broadway. The committee assigned to the negotiations was in favor of the purchase of this lot, and it was purchased from ex-mayor Richard Varick at the price of $10,680. At the next meeting, the committee presented the deed, and the bond and mortgage were executed. As notated in The Society's minutes, the following committee was appointed to take charge of the erection of the Mechanics Hall:

John Sherred, painter and glazer, member 1791; John McComb Jr., architect New City Hall, member 1791; Anthony Steenbeck, master mason, member 1792; Joseph Newton, member 1790; John Slidell, Jr., printer, member 1794; George Gosman, member 1791; Matthew L. Davis, member 1799; Andrew Morrell, member 1798; George Ireland, member 1796; Peter H. Wendover, member 1792; Cornelius Crygier Jr., member 1792; George Knox, member 1798 and Thomas Stevenson, member 1789.

Soon the building committee was vigorously working to erect Mechanics Hall. Master masons John McComb Jr. and Anthony Steenbeck were appointed to build the hall. A report of the opening of Mechanics Hall describes the building as follows:

In the basement was a spacious kitchen: on the first floor, a large coffee room; on the second floor, ceiling sixteen feet high, a large hall 52 by 25

> *feet, with a handsome orchestra and a drawing room twenty feet square. On the third floor were five spacious rooms for the use of clubs and meetings of any kind and on the fourth, twelve bedrooms...It soon became a popular place for balls and concerts.*

The cornerstone was laid on August 2, 1802, by the president of The Society, Society member (1792) Abraham Labagh. It was a fine brick building, and its front was embellished with the Hammer and Hand emblem. In the eyes of the public, it was a great ornament in the growing city.

The work completed, the building was dedicated on January 4, 1803, in the presence of the Honorable Edward Livingston, the mayor of the city, with

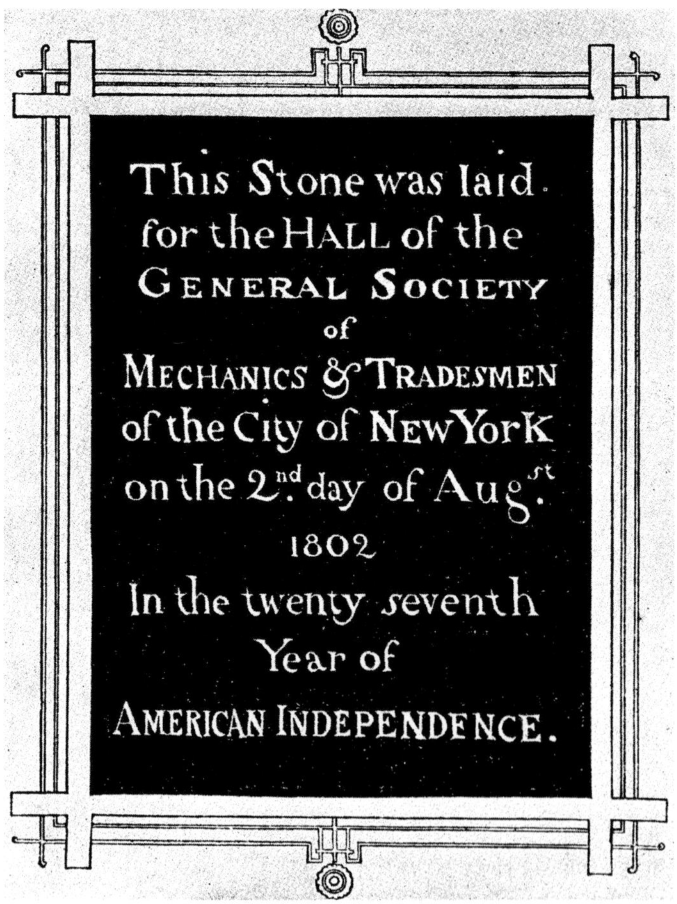

Cornerstone of the original Mechanics Hall as portrayed in the *Manual of the GSMT, 1916.*

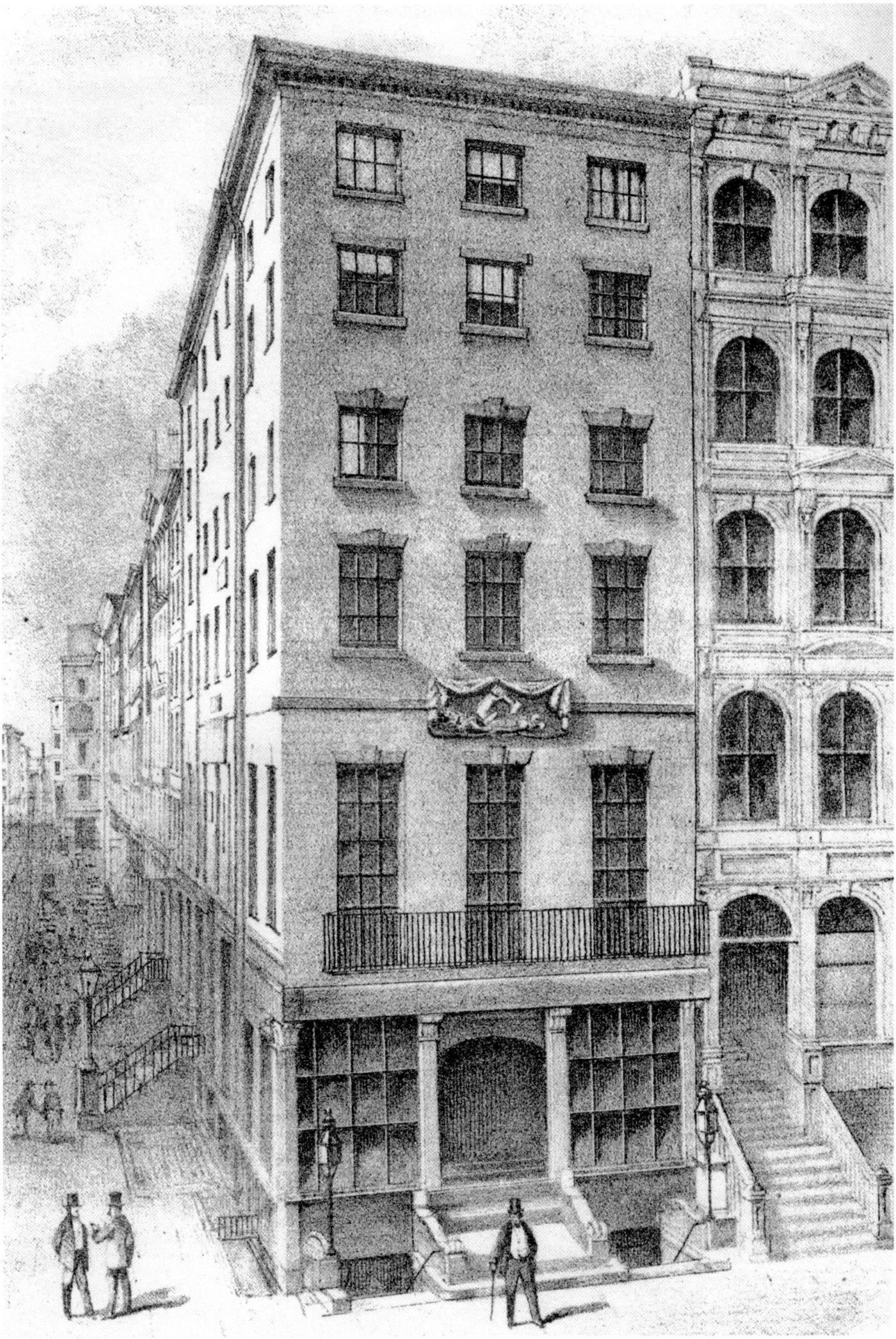

Original Mechanics Hall erected by The Society at 239 Broadway in 1802. *Manual of the GSMT, 1916.*

appropriate ceremonies and a large assemblage of distinguished citizens. At The Society meeting that evening, the following officers were elected for the ensuing year: Society member (1791) James Warner, president; Society member (1794) John Slidell Jr., vice president; Society member (1793) John P. Roome, second vice president; Society member (1790) John Stryker, treasurer; and Society member (1797) William Whitehead, secretary.

After The Society held its first meeting in its new Mechanics Hall, The Society, two hundred in number, assembled in another spacious room for the opening ceremony. A large turnout of distinguished citizens partook of the ceremony as well. The great celebration held, with sumptuous entertainment and many toasts celebrating Mechanics Hall, attests to the pride of The Society's members. In addition to the two hundred diners present on this memorable and brilliant occasion, Mayor Livingston and other dignitaries were in attendance. The *New York Gazette* and *General Advertiser* reported in their January 5, 1803 issues that upward of twenty toasts were responded to amid cheers and salutations including:

> *Toast VII: Mechanics Hall: May it stand firm as the principles of which it is founded.*
> *Toast X: May every honest patriot, beside an approving conscious, be blest with competency.*
> *Toast XV: Plenty: May it continue to smile on the labors of the American husbandman.*
> *Toast XII: Departed Worth: May the memory of Washington and his great compatriots excite to emulation, while it extorts tears of regret.*

Then, Mayor Livingston added:

> *The Mechanics of New York: May industry and skills, patriotism and integrity, be their characteristics.*

Avoiding Political Affiliation

The annual meeting of 1804 was held in the new hall with President John Slidell Jr. presiding, and much thought was given to the development of the aims of The Society in philanthropy and education. Rooms in the new Mechanics Hall were often leased for business purposes, and The Society decided to help in every possible way in civic and patriotic activities but to adhere to a steadfast decision to avoid political affiliations. As politics at this time greatly excited the community, it may be proper to note that, although the Federalist Party had its political headquarters at Mechanics Hall for many years, The Society itself ignored politics altogether. It had nothing to do with this occupancy, which was under an arrangement with the lessee of the hall.

Mechanics Hall remained in the same location until 1870 when a commercial iron building was erected, known as the Varick Building. The rental on this building became the source of a large part of The Society's income, the annual rental being nearly as much as the cost of the land and erection of the first Mechanics Hall.

During this early period, The General Society celebrated the mutuality and centrality of the craft community and played an important role in the festivities that marked patriotic holidays, carrying banners emblazoned with its motto, "By Hammer and Hand All Arts Do Stand."

A Centennial Celebration

Over the years, The Society took advantage of its longevity with award ceremonies, patriotic programs and noteworthy celebrations. By the time of its centennial celebration, the GSMT was a highly respected body known for its public contributions and its levelheaded approach to its responsibilities. As stated in the September 6, 1885 edition of the *New York Sunday Times*, The Society's "march of progress and usefulness has simply been prodigious. The General Society of Mechanics and Tradesmen shared, and helped bring about, the remarkable growth of a remarkable city."

The year 1885 was also a noteworthy one, as it marked The Society's centennial, which fittingly celebrated the 1785 founding of The General Society of Mechanics and Tradesmen with a splendid banquet, celebrated on November 16, 1885, at Delmonico's, the most brilliant dining place of the time. Heretofore, this kind of event was a male-dominated affair. However, this was the first time that the GSMT invited "the Ladies." The banquet was vividly described in the next morning's *New York World* as follows:

> *After a century of usefulness The General Society of Mechanics and Tradesmen sat down last night at Delmonico's to eat its Centennial dinner*

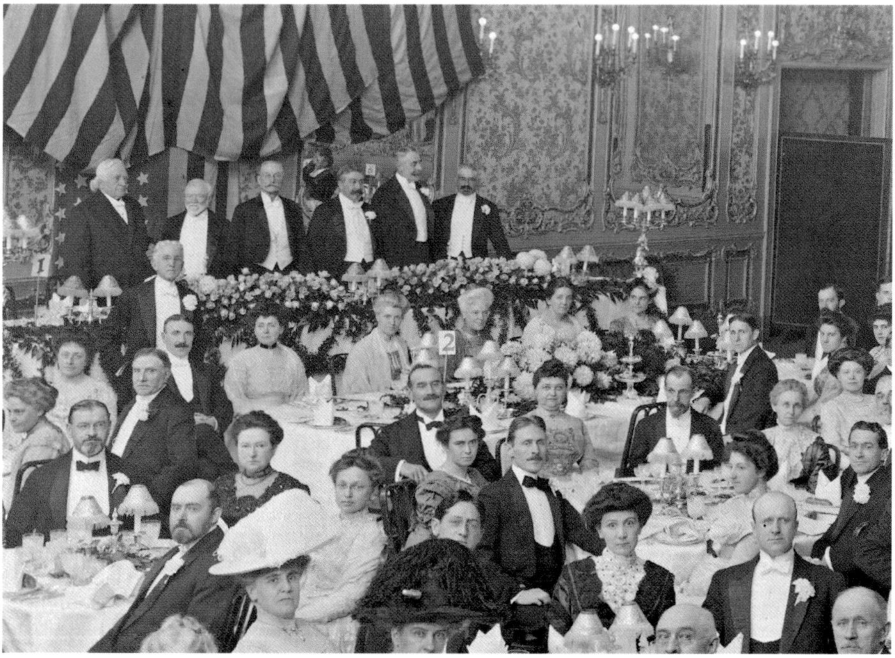

Delmonico's, the most the brilliant dining place of the era and the site of numerous GSMT celebrations. Note Andrew Carnegie on the dais in the background.

and listen to Centennial jokes. Not like the founders of this ancient Society did the feasters assemble. Not with rough, rousing choruses that shock the rafters of the Old Dutch taverns and made the fat bottles of schnapps dance on the shelves. Not with quaint queues and knee-breeches, and clay pipes with stems so long and so slender that the Lord only knows how they were kept whole towards the end of the meetings. There was no grandfather of Peter Cooper to head a committee and tell the landlady that drinks must be paid for on the spot, and no charge to The Society.

Above the heads of the radiant company which met last night hung great milky bulbs of Bohemian glass brimming with electric light, which sparkled from walls thick with gorgeous gold tracery. Instead of the venerable choruses howled out of tune, there was an orchestra in a yellow gallery, from which came the tremblingly "The Flowers that bloom in the Spring" and "Titwillow." Soft perfumes came creeping over the tables from hills and valleys of roses and violets. A forest of candles mellowed the upper lights, and a grove of great lamps, with silken shades bloomed in the middle of the room. Then there were the massive mirrors reflecting a sea of happy faces. But the feature of the occasion was the presence of so many ladies, the wives and sisters, daughters and sweethearts of the gentlemen.

Around the seven long tables were grouped the unostentatious representatives of industries that were the backbone and the heart of the city in all its stages of development. At the head of the hall sat in the chair of honor Chief Justice Charles P. Daly, with his white beard flowing to his waist, and on either side of him General Society President John H. Waydell and Senator Joseph R. Hawley. Flanking them were the Rev. Dr. Collyer, John T. Hoffman, Charles A. Dana, Judge John R. Brady, Abram S. Hewitt, Gen. Stewart L. Woodford., William H. Gedney, Whitelaw Reid, Chauncey M. Depew, Gen. Horace Porter, Ballard Smith and William H. Webb.

Down at other tables sat jolly old Forsyth Labagh, who joined The Society in 1828, C.H. Delamater, the iron founder, and his wife; Richard Poillon, the shipbuilder, and his wife; John Abendroth; Thomas E. Tripler and his wife and daughter; W. Otis Munroe and his wife and daughter; Thomas McElrath, John T. Agnew, Douglas Taylor and his wife; Andrew Little, Wade B. Worrall, Georg W. Lithgow and his wife and daughter; Hervey C. Calkin and his wife and daughter; Henry L. Slote and his wife; Alexander Knox, Charles T. Galloway and his wife and daughter; Robert Rutter and his wife; and Commissioner John R. Voorhis. There were men who had carved out great fortunes from the humblest and hardest beginnings on all sides. The whole scene was a tribute to honest labor, from the Rev. Dr. Collyer, who started as a blacksmith. It was a fitting end to the first century of this Society.

The General Society of Mechanics and Tradesmen records many "firsts" in its historical archives. For instance, the "Ladies" in banquet celebrations brought The Society notoriety, as it was the first time that a fraternal organization had invited women to its banquet, and soon other organizations followed the GSMT's lead.

The Mechanics Bank

The Mechanics Bank remains quite a noble achievement in the annals of the GSMT's history. From the early nineteenth century, mechanics and tradesmen sought opportunities to directly participate in banking and the GSMT's solution for a dependable source of capital was a bank owned and operated by mechanics.

In 1810, The Society took a giant step forward to reach its goal when delegates from The General Society were sent to Albany to present a petition to the state legislature for a bank charter. The incentive was to benefit mechanical interests and encourage manufacturing activities in New York City.

At a meeting held on March 28, 1810, the bank committee of the GSMT reported that the delegates to Albany, Gabriel Furman and Thomas R. Mercein, had returned triumphant with a certified copy of an act establishing the Mechanics Bank of New York. Previous to granting this charter, there were three New York banks doing business in the city—the Bank of New York, chartered in 1784; the Manhattan Company, chartered in 1799; and the Merchants Bank, chartered in 1805. In addition, a branch of the United States Bank, chartered in 1791, was also operating in the city.

The Mechanics Bank, the fourth New York–specific bank, obtained its charter because the mechanic interest was to be benefited by it and because the petition for it was endorsed by The General Society of Mechanics and Tradesmen. The lack of opposition was in part due to the craftsmen's considerable political clout and the strong support of the tradesmen. Society member (1792) Peter H. Wendover—president (1819), who was also a leading member of the Common Council of New York City—was largely responsible for obtaining the charter. The *American Citizen* publication, on March 2, 6 and 8, 1810, discussed the Mechanics Bank's raison d'être:

> *Supporters of the Mechanics Bank described the city's need of and ability to absorb new capital; the importance of an alternative to the "capricious, irregular and arbitrary manner" in which local banks operated; and, finally, the want of an institution that would be orientated to the problems of tradesmen, such as discounting notes of small denomination with which other banks would not bother.*

Society member (1799) Matthew Davis, a politician and former printer, took it upon himself to write the following to a friend:

> *It is impossible for you to imagine how popular the measure is; and you would smile to see the zeal displayed on the subject by mechanics of every grade whether Madisonian, Clintonian Federal or anything else that you can designate.*

From 1810 to 1837, the Mechanics Bank became one of the largest in the United States. It aided GSMT members by loaning money at low interest rates for The Society's various activities, including real estate investment, a

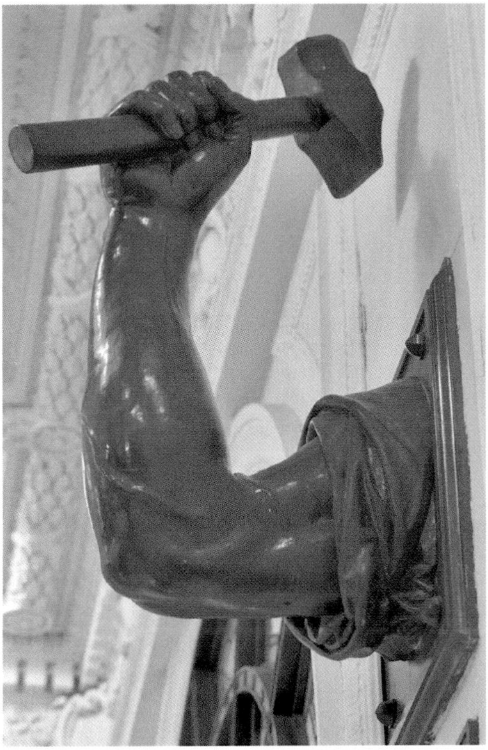

The *Hammer and Hand* bronze sculpture originally hung over the doorway of the old Mechanics Bank at 33 Wall Street; it's now mounted in The Society's lobby. *Courtesy of Stephen Amiaga.*

school, a library and the erection of halls and buildings.

The capitalization of the bank was for $1,500,000. The legislature would subscribe to $250,000 worth of shares, and $600,000 worth of shares was reserved for the mechanics of the state.

At a subsequent meeting, the president of The General Society, George Ireland—who was a member (1796), vice president (1809) and president (1810)—reported that the president of The Society is always to be, ex officio, a director of the bank and that seven other directors must be chosen from among the members of The Society, four of whom must be engaged in some mechanical profession and that the state may provide, by law, for the appointment of two other directors—the effect being to place the control of the bank with The Society.

Some of the first directors of the bank were Society members:

Jacob Sherred, a painter and glazier, Society member (1791) and president of The Society (1796, 1801).
Francis Cooper, a blacksmith, member (1793) and second vice-president of The Society (1802).
Stephen Allen, a sailmaker, member (1792), second vice president (1806), vice president (1807), president (1809).
Matthew L. Davis, an insurance man, Society member (1799).
Samuel St. John, a tailor and merchant, Society member (1792).
Gabriel Furman, joined The Society (1791) president (1795).

When the Mechanics Bank's charter, which was limited, expired, it was renewed upon application of The Society. For a number of years, The Society was well represented on the board of the bank until the duties became too onerous for busy mechanics.

The Society then chose their eminent member John Slidell Jr.—a prominent soap boiler who joined The Society in 1794—as the first president of the bank. He had circulated a petition for a Mechanics Bank as early as 1805. Slidell had a strong commitment to The General Society and had served as its vice president in 1803, president in 1804 and its treasurer from 1809 to 1825. Former Revolutionary War officer and New York politician Nicholas Fish, a non-member of The Society but an esteemed financial wizard, was the bank's first cashier. Fish was the father of Hamilton Fish, an American statesman and politician and sixteenth governor of New York.

The bank had its beginnings in a three-story building in which Alexander Hamilton had previously made his home. The offices opened on the ground floor, and the president of the bank and his family lived over the counting room. Within two years of operation, the Mechanics Bank had become one of the largest and strongest banks in the United States.

The year 1812 brought rumors of war, and The Society decided to dispose of some of the bank stock, which was sold at a handsome profit. By the early 1820s, The Society was expanding into different fields. It financed those ventures, in part, with a loan from the Mechanics Bank and the rest from the treasury of The Society, the balance open for voluntary contributions from members and the general public. Within seven years, the Mechanics Bank had grown and prospered to the extent that the federal government put one-fifth of its deposits into its hands.

Decline of the Mechanics Bank

Despite the Mechanics Bank's aim to provide artisans with liberal credit and convenient banking services, elements of discontent surfaced among the membership.

According to the minutes of The General Society for January 19, 1823, the Mechanics Society's membership had changed. It had considerably loosened its membership requirements and approved a great number of well-to-do merchants whose only connection was that, at some time in their lives, they had followed an artisan trade or been connected with the business operation of a craftsman.

A number of these members, such as real estate broker Samuel St. John and insurance broker Gabriel Furman, became members of the board of directors because they were familiar with banking operations, as genuine artisans such as Stephen Allen, Society president in 1809 and also a director, were not. As a result, St. John and Furman managed to assume the overall management of the institution, including the determination of who would and would not receive stocks and what notes were to be discounted. Shares were sold to those who would keep the best account of the bank rather than to the artisan community as the charter intended. Forced to stop payments, the bank soon became insolvent.

The Mechanics Bank subsequently never fulfilled its promise nor satisfied the credit needs and financial services that artisans sought. The General Society sold five thousand of its six thousand shares, divorcing it from any real power over the bank. The minutes of The Society for December 13, 1810, acknowledged the reason for discontent:

> *The problem was our Brothers who have been Directors of the Mechanics Bank, who were guilty of very unfair and improper conduct particularly in the distribution of that stock which was by charter reserved for mechanics.*

The Mechanics Bank barely survived the cacophony of discontent; however, it was too late to restore the trust and leadership previously afforded The Society by the city's mechanics. In summation, undoubtedly, the Mechanics Bank's demise was attributed to the tradesmen's inexperience with such complex institutions as commercial banks. However, there were other factors that contributed to the bank's decline.

In 1832, there was a cholera epidemic that caused a great deal of loss, and in the same year, violent riots resulted in considerable property damage. New York seemed at the brink of destruction when, on December 16, 1835, the great New York fire destroyed a total of 648 buildings in which $18 million of real estate was destroyed. The bank did not escape the consequences for its and The Society's reliance on investing a large part of their assets in real estate. Another ominous factor came with the specie circular issued in July 1836 and the bankruptcies and business depression of May 1837 when all the banks in the country stopped paying out specie.

The final blow occurred on May 4, 1837, when John Fleming, the bank's ex-president and chief cashier, died unexpectedly. Many depositors believed that he had committed suicide because of the poor financial condition of the bank. Fearing the worst, this event—coupled with other rumors and the general hysteria—resulted in a great fall in the price of the bank's stock. The lack of faith instilled in the depositors and the panic-induced selling only added to the uncertain condition of the bank during the Panic of 1837.

After that, although the bank attempted to regain the support of The Society, it was unable to do so, and once that happened, the bank could no longer maintain its position of supremacy in the financial world. It continued under various other names, such as the Mechanics and Metals National Bank, and was finally incorporated into the Chase National Bank.

At the time of the Mechanics Bank's demise, The General Society of Mechanics and Tradesmen was presented with the emblematic bronze Hammer and Hand fixture, which for years was over the doorway of the old Mechanics Bank at 33 Wall Street. Today, this imposing sculpture is mounted on a wall in the lobby at the GSMT, 20 West Forty-fourth Street. A similar brass plaque is also mounted on the outside of the building.

8

THE MECHANICS SCHOOL AND THE MECHANICS INSTITUTE

Education in a free and an enlightened society offered even the humblest artisan the chance of becoming the next Franklin, either as scientist or patriot.
—Thomas Royce Mercein

As early as 1802—at a meeting held in the GSMT's first hall, located at 239 Broadway—the education mission of The General Society came to the fore of discussion, and a school to provide free education to children of needy members was on the table. At this time, free education had been established by charities and benevolent organizations, and nearly every church had a school. However, a public school system did not exist in New York, except one in the almshouse and the other open to the children of freed slaves.

Lacking a public school system, The General Society was deeply concerned about the education of its members' children and began to focus on free education for children and how much of The Society's revenue might be appropriated for this project.

With their visionary outlook, the members convened on this subject. Brother Thomas Royce Mercein was a strong advocate for education. Like the other members, he believed that education cleared a path for the advancement of all able men, including thousands who would later attend the Mechanics Institute. A master baker, Mercein was a Society member (1803), who moved up the ranks to become second vice president (1814), vice president (1815) and president of the GSMT (1827). Throughout his tenure, Mercein played a pivotal role in establishing the school for the children of its members.

The Day School

Free Education

A school to provide free education took on a stronger appeal when The General Society's funds were sufficiently sound to consider free education for the members' children. As noted, the educational facilities already in existence were inadequate to meet The Society's needs. In addition, other schools established by the more affluent class charged high tuition fees that many of the members of The Society could not afford nor could the children of indigent and deceased members secure an education.

However, other issues must have taken precedence at The General Society meetings because interest in the education project seems to have waned until 1819 when a new education committee was appointed. At a General Society meeting on March 1, 1820, the committee in charge of education of the children of indigent members reported:

> *The committee on education presented a plan for the establishment and maintenance of a school for the children of members of The Society and the report was accepted and the committee was directed to carry out its suggestions and to find suitable rooms for the school, which could also accommodate a library. Twelve members were chosen as the school committee to employ teachers, make regulations and visit classes periodically.*

The report also stated that the education plan would also include children of deceased members in good standing whose mothers could not afford to pay for their education and would admit outsiders for a modest fee. In the same year, members of The General Society founded two educational institutions: the Day School and the Apprentices' Library, which was founded as an educational resource for apprentices and factory workers. The Library is the second-oldest library in New York City. The GSMT's reasons for setting up the school were given as follows:

> *The course of education in our Mechanics School enables the young mechanic to acquire a knowledge of reading, writing, and the rules of common arithmetic; but beyond these branches it scarcely makes any pretentions, he therefore labors under a serious difficulty at the very commencement of his business, namely that of being compelled to learn a set of dry and often uninteresting rules, without a previous acquaintance with the principles of which they are founded.*

It was with the view of supplying this deficiency, as well as others thereafter to be mentioned that the school was established. It was designed as a school for teaching the most useful branches of physical and chemical science, to prepare the mechanic to understand and appreciate the lectures of the college or university, and thus increase his knowledge and usefulness and happiness.

The Mechanics School

In 1820, rooms were taken in the New York Free School Building in Chatham Street, a teacher was hired and the school opened with seventy students. At the same time, the gift of a thousand books made possible the opening of the Apprentices' Library.

However, The Society wanted a schoolhouse and library of its own, and a year later, on June 13, 1821, with great ceremony, the cornerstone was laid on a plot of land leased from the city for sixty years at 10–14 Chambers Street. The cornerstone of the building was dedicated on November 26 of the same year with suitable exercises. In popular reference, the school was referred to as the Mechanics School, which remained at this location until 1823.

At the opening ceremony, prominent Society member Mordecai M. Noah, then the president of the *New York Advertiser*, delivered the dedication address. He was initiated as a member in 1821, the same year the school's cornerstone was laid. Noah was a strong supporter of the Mechanics School's role in the development of young men who he predicted would become pillars of society.

Student Qualifications

To enter the school, children had to be fourteen years of age or younger and be able to "spell words of two syllables." According to the school committee's minutes:

Examinations were being given in geography, grammar and arithmetic. In 1821, the number of scholars was 95 (55 attended free of charge). As of December 21, 1822, there were 105 scholars (56 attended for free). The subjects mentioned in the minutes are "reading, writing, arithmetic, grammar, geographies, with the principle use of the globes and maps and the elementary principles of astronomy."

The school soon became very popular and, through a greater part of its existence, was self-supporting. As enrollment to the school continued to grow, the committee focused on establishing a separate school for girls.

By January 4, 1826, there were 120 scholars, and the minutes noted discussion regarding a separate school for females. The female school opened May 1, 1827, with 75 pupils, 19 taught free. Pupils were "daughters of the most respectable and valuable members of The Society." It was among the first schools in the city that concerned itself with the education of females. The studies were nearly similar to those in the boy's department, except for needlework ("plain and ornamental") and painting and drawing.

THE MECHANICS SCHOOL'S NEW QUARTERS

The Society moved its school many times, each time to larger quarters. In 1831, at the end of a dozen years, the Chambers Street building was found inadequate for the school and library. The Society, with its vision for expansion, purchased a three-story brick school building and moved to 30–36 Crosby Street.

However, the school's property so quickly founded was temporarily interrupted by a scourge of cholera in 1832, and with permission of The Society, it caused the building to be occupied as a hospital. Improvements to that property were suspended until about 1834.

During that year, the building was repaired and fitted up. The library, the school and meeting rooms of The Society were then reestablished in the Chambers Street building. On the road to recovery, the school again became self-supporting, with about one-eighth of its pupils being taught gratuitously. When the school reopened, it had 244 pupils in the male department and 197 pupils in the female department, with Arabella Clark as principal with five assistants. The courses taught also expanded into new fields of study:

By 1835 the school began to offer classes in architecture and engineering in the boy's school. In addition to a class on "mensuration of heights and distances, plain and spherical trigonometry, and land surveying," extra

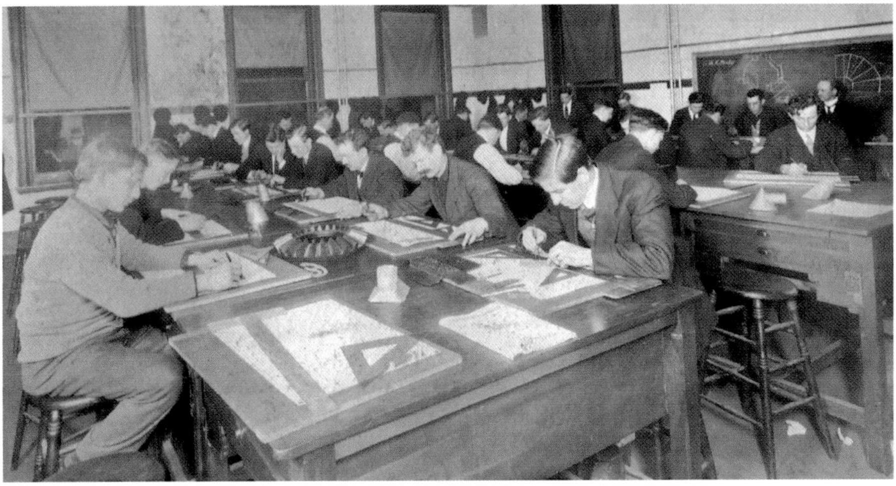

Young men in studious pursuit at the Mechanics Institute.

> *classes were offered in conic sections, architecture, mechanics, chemistry, mechanical philosophy, drawing and designing, and civil engineering.*

Scholarships

In 1835, Columbia College extended the privilege of free scholarships to the Mechanics School and offered to accept six scholars free of charge annually. At the same time, Professor Anthon, rector of the grammar school, agreed to have always six scholars from the Mechanics School in a course of successive preparation for college. In 1840, the Honorable Theodore Frelinghuysen, chancellor of the University of the City of New York (New York University) announced that the Mechanics School would be allowed to send two scholars to the university and not have to pay tuition. These opportunities, The Society hoped, would allow the mechanic to take his seat "in the Assembly, in Congress, or in the Cabinet, with all good, sound, common sense more available to that service than the attainments of the bar, the pulpit or the material media." In an address before The Society in 1850, Mordecai M. North said:

> *How are we to account for the rapid progress of our country to arts, civilization, literature, commerce, and science? By our free institutions, the*

> *quality of our laws, and, above all, by the free education which visits all alike, from the cottage to the palace. What has led to this prosperous current which is carrying us ahead of all our sister states and cities? Our commerce on one hand, and the enterprise and the laborious industry of our mechanics on the other. The time has arrived when it has become apparent that the destinies of our country are placed under the control of the mechanics and laboring men. Well-educated mechanics will fill our legislature and the hall of congress; their numerical strength will accumulate until they are able to command the highest stations in the government.*

Noah continued his long and eloquent speech:

> *What will be our guarantee for the safety of the country? I answer in the education and intelligence of this class of our citizens.*

Noah further cited the example of Millard Fillmore, who was elected to fulfill the presidential chair through the death of President Taylor in 1850:

> *At an early age he was sent from home to learn the clothier's trade, and about four months later was apprenticed to a wood-carder in the town where his father lived. The President of the United States was a mechanic—an apprentice boy, as many of my bearers have been. True in after life he studied law, and was a successful practitioner; but he carried into that study and into that practice and into the high station he now adorns the elements of patient industry acquired as an apprentice.*

During all this time, the philanthropic aid of The Society continued to be administrated with care and thoughtfulness, including pensions and help to the indigent.

THE FEMALE DEPARTMENT

New York's population had grown to about 1.25 million, and the city was taking its place as the business and money center of the world. The free school was flourishing and proving itself to be of inestimable value to apprentices and the young women who were also receiving instruction at the school.

It is significant to note that one area in which the female school body excelled was as teachers. So superior was its course of instruction that its graduates were in demand as teachers for other schools. Attesting to this career direction, in 1841, it was reported that forty of the young women who had graduated became teachers in public and other schools in New York City.

The Mechanics Institute Evening School

In 1858, the tides were turning to provide better education for the masses. At this time, New York City established its public school system, and the GSMT decided that the daytime Mechanics School had fulfilled its mission and that the increasing merits of the public schools made it unnecessary to continue the work begun at a time when the city school system was undeveloped. At this time, the Mechanics School became a night school for older students who worked during the day, eventually taking the name "Mechanics Institute." The report of the school committee on March 3, 1858, explains The Society's decision:

> *The circumstances by which the school was surrounded have changed; the public schools have greatly increased in numbers and improved in character; many of them are now superior, or at least equal, to the best private schools; parents no longer deem it derogatory to their children to participate in the advantages of those free institutions which they are so largely taxed to support and maintain. The School is no longer a necessity: it has fulfilled its mission; its work is done; for nearly forty years we have cherished and sustained it, an honor to our Society, a blessing to its members and their children, and an important public benefit.*

Thus, The General Society officially closed its day school, and a tuition-free evening school, The Mechanics Institute, began offering technical training to young men, consisting of free-hand mechanical and architectural drawing, modeling in clay, elementary mathematics, technical lectures and material aids.

At the inaugural address of the Mechanics Institute, Society member (1847) Ira Hutchinson, president of The Society (1858), commented:

This Society has done much toward removing the prejudices and disadvantages under which the mechanics of this city have labored. But the great work of redeeming the laborer from frowns, and his elevation to a standard commensurate has scarcely yet begun. There is no class of men, nor any association, so abundantly qualified in means and abilities as The Mechanics and Tradesmen's Society of the City of New York; and there is no class of men among the inquirers into the principles of nature that the world can look to with more assurance for usefulness than to the true mechanic; and to no class among the rising generation could our efforts be more advantageously applied than to the apprentice in the accomplishment of this great work.

LARGER QUARTERS

The ever-increasing need for a larger building to house The Society's Mechanics Institute became pressing. In June 1877, The Society purchased the spacious Suydam mansion at 18 East Sixteenth Street, The Society's fourth location. After the necessary alterations and furnishing, the Mechanics Institute, library reading rooms and business offices were removed from the Broadway and Crosby Street buildings, which had been leased for business purposes. As tradition would have it, the new building was dedicated with appropriate exercises, and the opening reception took place on the evening of January 22, 1878, from seven to ten o'clock. The weather was mild and pleasant; the visitors numbered nearly six hundred.

In 1879, John R. Voorhis became president of The Society. He had taken very active interest in the establishment and financing of the new quarters on East Sixteenth Street. In his inaugural address, he expressed his views:

I trust our efforts will not rest here, but that we will continue to foster and enlarge this worthy feature of our Society, until it shall be its privilege and duty to occupy a location, with a building thereon inferior in its design, construction and adaptation to its purposes and needs to no other in America, from and through which it will diffuse on every direction its benefits to all the industrious and producing classes of this vast city, taking its proper position, and receiving the deserved recognition of the leading Society of this great metropolis.

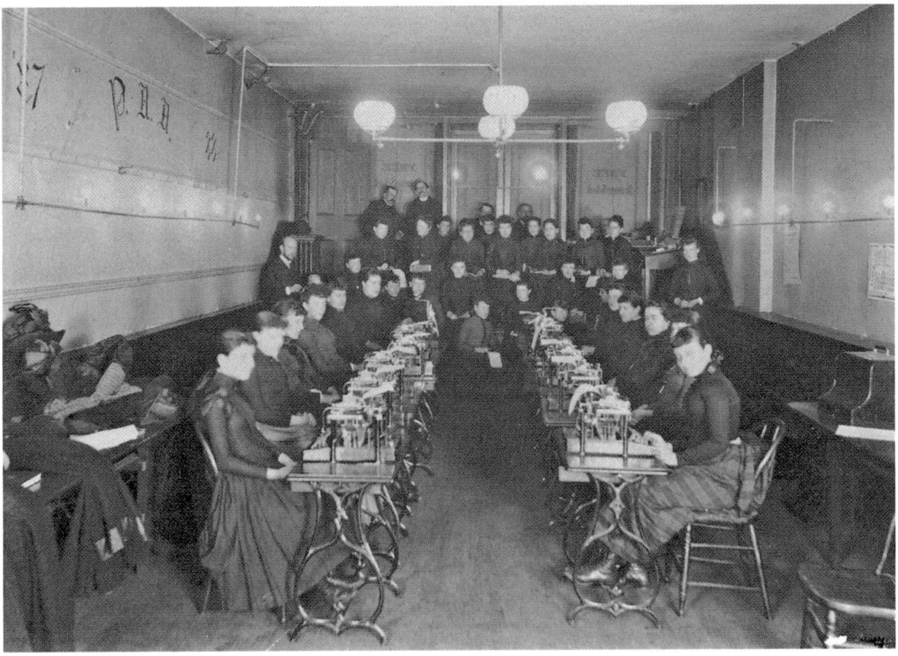

Young women at their typewriters at the Mechanics Institute.

In 1877, courses in stenography and typewriting were added to the Mechanics Institute's offerings. Many of the young women came to the school tired and discouraged from arduous tasks, the uncongenial surroundings and health-destroying atmosphere of the factory, yet they entered the classroom with a zeal and enthusiasm in marked contrast to their physical conditions. The records in the GSMT archives give gratifying evidence that these women accessed a higher plan in their social conditions. A young woman wrote:

> *I wish to state that I feel very grateful to the members of your society for all they have done in my behalf. Before I studied stenography and typewriting I worked at dressmaking, with long hours and very small wages. But thanks to the instruction I received in your school I am able to fill the position of stenographer and typewriter in a law office where my work is very pleasant and my salary remunerative.*

Another student wrote:

> *I have now had my position for little over two months. I want to thank the society, for enabling me to obtain such agreeable occupation. I am glad to*

> *add one small link to the long and strong chain of evidence of the help that The General Society of Mechanics and Tradesmen have extended to girls, in first facilitating the study of stenography and typewriting, and afterward procuring both lucrative and congenial situations.*

By 1894, the architectural class was composed of masons, carpenters, stonecutters, ornamentors and bricklayers in all degrees of proficiency. The class in cabinetwork and decorative design composed of cabinetmakers, furniture designers, decorators and wood carvers. The mechanical class was a preparatory class including the mechanical drawings required by stair builders, metalworkers, machinists and pattern makers. A machinist in the shop of the Standard Oil Company confirmed:

> *That this instruction enabled him to prepare working drawings for important machinery, and is much grateful at his success, giving credit to the school, therefore.*

At the same time, a journeyman mason had been advanced to position of foreman by his employer, owing to his knowledge of plans and drawings for masonry acquired in the architectural class. A grateful student wrote:

> *You may be interested in my Alma mater, Mechanics Institute. They accept students regardless of age, providing the student is working at a job connected with the course. I graduated at the age of 50. So you can understand my enthusiasm for the school, especially when there are no tuition fees.*

THE GSMT'S HEADQUARTERS

As the requirements for enlarged educational facilities for The Society's work became apparent, The Society members decided that another building was needed. Seeking larger quarters, the GSMT purchased property at Seventh Avenue and Forty-eighth Street. Before a suitable building was erected, the substantial structure at 16–24 West Forty-fourth Street was found to be better adapted to the requirements of The Society. The address was later condensed to 20 West Forty-fourth Street.

After negotiations with the officials of Columbia University, its then owner, The Society purchased the property in 1899. It is interesting that

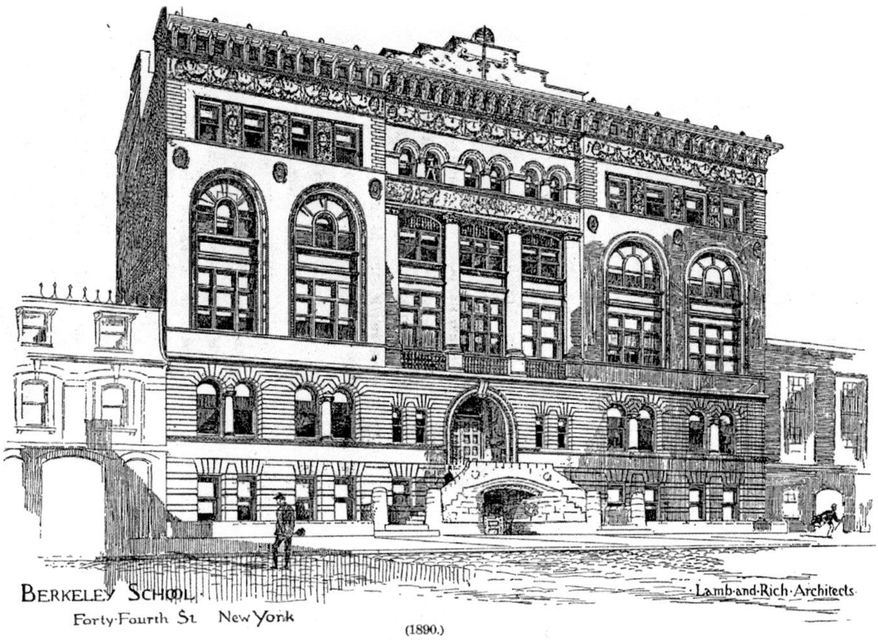

Historical image of The General Society of Mechanics and Tradesmen of the City of New York, formerly the Berkeley School. The Society purchased the property in 1899.

the architectural firm Lamb & Rich, one of the city's most prominent and prolific architectural firms in the last quarter of the nineteenth century, originally designed the building for The Berkeley School for Boys in 1890. Military service was required of the young men, and the ground floor drill hall is still intact, though it is now renovated and referred to as The General Society Library. The Library's main reading room soars three stories and is topped by a magnificent skylight. The building is the fifth and current home of The Society, which was dedicated with appropriate fanfare to its new uses on January 3, 1900.

The Society was solvent, and generous gift contributions came in from prominent members. Andrew Carnegie, for one, is best remembered for contributing over half a million dollars to the GSMT, which enabled The Society to greatly extend its work so that the new building was placed in condition for occupancy.

In 1903, Carnegie's gift of $250,000 made possible the addition of two floors, known as the Carnegie Extension. The generous funds made possible

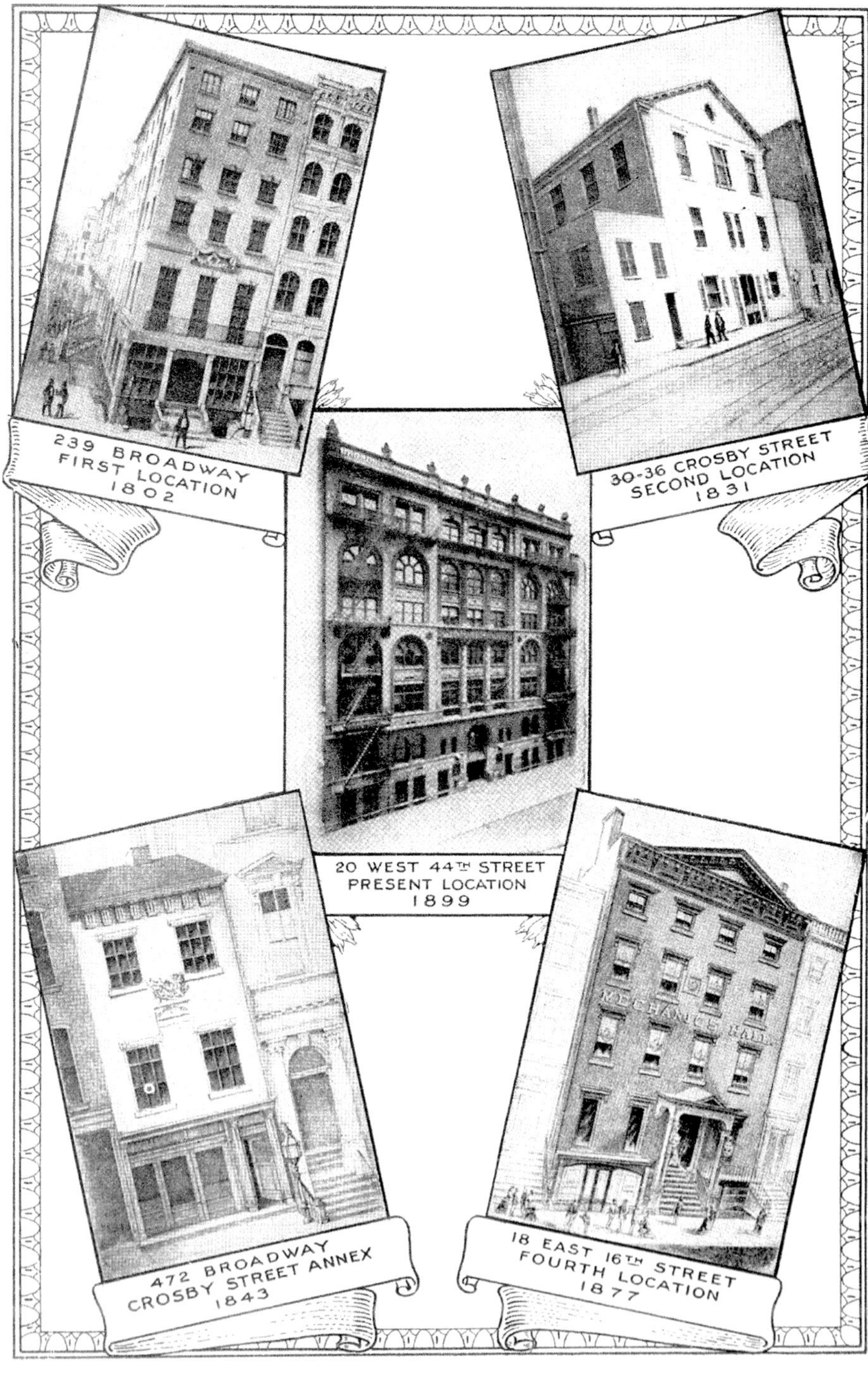
239 BROADWAY
FIRST LOCATION
1802
30-36 CROSBY STREET
SECOND LOCATION
1831
20 WEST 44TH STREET
PRESENT LOCATION
1899
472 BROADWAY
CROSBY STREET ANNEX
1843
18 EAST 16TH STREET
FOURTH LOCATION
1877

the extensive renovation and expansion of the building. Society member (1905) Ralph Samuel Townsend, the architect, worked on the project, which included the additional floors and classroom wings for the Mechanics Institute and the formation of an impressive reading room, from 1903 to 1905.

In 1908, Carnegie gave an additional $200,000 as an endowment fund, the interest of which was to be used for the educational purposes of The Society. Thus, The Society was enriched, $527,000 in all, by this generous benefactor. At the time of the building's opening, Carnegie was an active and honored Society member (1891), yet he refrained from acknowledgements and accolades. He wrote:

> *I shall certainly visit the rooms very soon. I do not wish to be lionized and therefore stayed away from the ceremony. There is not necessity for a committee thanking me, because I consider it a great privilege to be permitted to aid The Society, and I beg to thank the committee for allowing me to do so.*

To summarize: The General Society purchased its first building in 1802 at 239 Broadway and called it the Mechanics Hall. Then in 1831, The Society's second location was at 30–36 Crosby Street. The third location, which The Society moved to in 1843, at 472 Broadway was the Crosby Street Annex. The fourth location, first occupied in 1877, was located at 18 East Sixteenth Street. The center image in the illustration shows the GSMT's present location at 20 West Forty-fourth Street, the fifth and final building, purchased in 1899.

Before the twentieth century dawned, The Society was established in this splendid building. Its permanent headquarters, in a single structure at 20 West Forty-fourth Street, houses the GSMT's Mechanics Institute, The General Society Library and cultural programs, now being carried on in the heart of the city. The façade of the building is a New York City landmark, and the building itself is listed on the National Register of Historic Places.

Home at last in their permanent headquarters, The Society, since its founding in 1785 to 2015, is a 230-year-old institution that continues on a steady course of public service to the citizens of New York City in the fields of education, lectures and literature. By preserving this historic structure for posterity the GSMT salutes the Brothers whose wise real estate investments made it possible.

Opposite: Montage of the GSMT's five building locations from 1802 to 1898. The center image represents The Society's fifth and final location at 20 West Forty-fourth Street.

THE MECHANICS INSTITUTE TODAY

The General Society's Mechanics Institute continues in its solitary role as the oldest tuition-free evening school in New York City and the only remaining mechanics institute in the country with a continuously operating technical school and library services.

It has been providing job training since it was established in 1858 to qualified students in crafts and building trades. In keeping with the times, the Institute currently provides instruction in two- or three-year programs that are taught by instructors who have hands-on experience in their trades and have a sincere desire to pass on their experience and knowledge to others. The Mechanics Institute states:

> *It is for this reason that our Mechanics Institute had long been recognized as a traditional meeting ground where men and women can develop and enhance their abilities under the guidance of accomplished instructors who work in the field. The Mechanics Institute enrolls all applicants meeting the entrance requirements, regardless of race, color, creed, sex, age, or place of national origin.*

Complete requirements for qualification and registration are available in the Institute's catalog. Of the eight awards and scholarships of significant historical note is the George E. Hoe Prize Fund, which is awarded to graduates who have demonstrated significant achievement during their enrollment in the Mechanics Institute.

In 1858, young George Hoe won a prize in the Mechanics Institute and this recognition gave him such a lift that, thirty-eight years later, in 1896, when he became The General Society's president, he donated $500 for a student prize and kept adding to this until he had given a total of $15,000.

A number of other gifts came through the years from members and well wishers and by those graduates who got their start toward success thanks to The General Society's free instruction. On a steady basis, many other gifts, both large and small, enabled The Society to move forward with the times.

While the Mechanics Institute is not a trade school, it does offer comprehensive instruction in a wide variety of subject areas related to the building and construction industry and has trained more workers in these fields than any other institution in the nation.

A Landmark Façade

When you first visit the GSMT building on Forty-fourth Street that houses the Mechanics Institute, your attention is drawn to the building's landmark façade, which has been well preserved. It is a building filled with history and rich with tradition. It remains a fitting symbol of the great contribution the GSMT has made to this great city.

As you enter the building, you find yourself facing The Library's oak, brass and glass doors. To the right is an elevator, which provides students and visitors access to the building's six floors. To the left is a sweeping marble staircase, fitted with polished brass handrails and amber-shaded lamps. Just below the staircase on the wall is the cornerstone commemorating the original Mechanics Hall, inscribed:

> *This Stone was Laid for the Hall of The General Society of Mechanics and Tradesmen of the City of New York on the 2nd day of August 1802 In the twenty-seventh Year of American Independence.*

The Institute's administrative offices are located on the second floor, and the building also houses the computer labs, hands-on workshops, lecture classrooms and a faculty lounge.

Outreach Programs

In recent years, the Mechanics Institute has also developed and conducted client-tailored instruction workshops for various business entities within the building and construction trades industry. Past industry clients include the New York State Department of Correctional Services, New York City Department of Buildings, the New York Public Library, the DC-37 Educational Fund, the Regional Alliance for Small Contractors, Consolidated Edison of N.Y., the International Alliance of Theatrical Stage Employees (Local 1, IATSE) and the Insulation Contractors Association of the City of New York.

The General Society of Mechanics & Tradesmen of the City of New York

Mission Statement

The Mechanics Institute's historic existence is founded on the belief that education is the key to life's success.

> *Our steadfast conviction is built on our commitment to help those who are willing to help themselves. To this end we are dedicated to:*
> *Preparing the foundation for those willing to rise;*
> *Providing the ladder for those willing to climb; and*
> *Fueling the spirits of those willing to soar.*

The GSMT's Mechanics Institute is not a trade school but holds coeducational technical classes where individuals who have learned or are learning their trades can obtain the necessary technical knowledge to help them advance in a large variety of fields such as construction, real estate, preservation and government.

Faculty members, like the students, work professionally during the day and bring knowledge from their fields of expertise to the classroom. Full-time students graduate after two or three years of instruction and receive a diploma. Other students take individual courses specific to their line of work.

Since 1858, more than 180,000 men and women have studied at the Mechanics Institute in their quest to advance in today's world. Today, the Mechanics Institute continues to provide tuition-free evening instruction in trades-related education and is a unique service to improve job opportunities for New Yorkers in the building and construction industry.

Throughout the twentieth century, The General Society's Mechanics Institute continued to provide vital educational opportunities, during both periods of prosperity and the years of the Great Depression. In 1931, The General Society reported that it had enrolled a record number of students, noting that:

> *It is significant in this time of economic depression, the opportunities extended by the society to young men to equip themselves to command a better livelihood and enjoy a higher standard of living, are more in demand than ever before.*

In 1942, The General Society conducted an evening school sponsored by the United States government that taught one thousand men from industrial plants drafting skills for the expansion of the production of munitions for the war. Courses taught included mechanical drafting, freehand drafting, blueprint reading, mathematics for machine shop practice and lettering and tracing for mechanical drafting. This effort continued into 1943 with the training of aviation mechanics.

A PROMINENT DIRECTOR

A young man, Society member (1934) Louis Horace Rouillion, whose ability had been brought to the attention of the school committee, was made director of the Mechanics Institute in 1957. In an interview in the *New York Daily Mirror*, in the "Only Human" column by Sidney Fields, Rouillion said:

> *Students range in age from seventeen to over sixty, said Rouillion, which rhymes with bullion. But 90 percent are between 25 and 30, married, and have children. The students take everything from a special course to a full three-year course, which wins them a diploma. We're not a trade school, no shops here, only classrooms. We're a technical Institute. It's trained more men for the building construction industry than any school in the country, though it is completely independent of business. It has $4,000,000 in endowments, faculty members have degrees in science and engineering, and like students, work during the day. My father Society Member (1907) Louis Rouillion came here as director in 1900, when it had 200 students. He built it to 4,000.*

During Louis Rouillion Sr.'s Mechanics Institute tenure, he garnered praise and appreciation. The following comments were recorded in a GSMT testimonial book attesting to Rouillion's contribution as director from 1900 to 1945. Students who attended the Mechanics Institute recall:

> *Andrew G. Hagstrom—Greeting from one of the thousands of former students who recognize you as The Mechanic Institute and whose privilege has been to know and respect you for 25 of the 35 years.*

> *Another student—For it was through Louis Rouillion, by whose indefatigable efforts, Mechanics Institute reached its present high position among technical schools. During the three and a half decades of Louis Rouillion's regime between 6 and 7000 students received the diploma of*

the Institute, and many members of the alumni have been received into The General Society.

Benjamin Kriegel—Without knowledge there is no sure progress.

GRADUATES REMEMBERED

Among the more famous alumni of the Mechanics Institute, Society member (1909) Andrew H. Dykes, the big lumberman of Dykes Lumber Company, was a member and secretary of the school committee for several terms and vice president (1933) and GSMT president (1934). Dykes received part of his education at the Mechanics Institute.

Andrew Hagstrom was a Swedish immigrant whose company, Hagstrom Map Company, made detailed maps of all the boroughs. The maps had a new, improved look: a dramatic diagram form, with smaller blocks, wider streets and numbers at street intersections. At first, he distributed the maps as advertising, but in time, he was amazed to find how rapidly the maps sold. Hagstrom honed his skills at the Mechanics Institute. He began evening classes in 1909 and continued for three years until he graduated and received his diploma. Later, he became an instructor at the Institute, where he taught for nine years. He is co-author of a textbook called *Construction of Lettering for Mechanical Drafting*, which is used in the Mechanics Institute.

Hagstrom had a long relationship with the GSMT. He was initiated a member of The General Society (1928), and served on the literary committee (1933–40). He was elected second vice president (1950), vice president (1951) and president (1952). In his inaugural address, Hagstrom said:

> *No other Society in the world would take a complete stranger, a nineteen-year-old immigrant boy from Sweden, who knew neither the language nor customs, give him a free education, admit him to your Society, and finally elect him to the highest office of The Society.*

Under the auspices of the Carriage Builders' National Association, Course B12: Carriage and Automobile Drafting was intended to teach men to design and make drawings and fashion plates for them. The class was limited to those employed in carriage and automobile building. Raymond

Dietrich, a Mechanics Institute student from 1915 to 1918, worked as a draftsman by day at Brewster & Company.

An ex-Brewster draftsman named John D. Gibbon founded this technical school for carriage draftsmen in 1880. In 1892, Andrew A. Johnson, another former Brewster employee, took over as instructor-in-chief, and it was commonly called the Andrew A. Johnson School in the Mechanics Institute.

The terms of Dietrich's apprenticeship to the school required a sponsor, and Willie Brewster, president of Brewster & Company, signed off for his eager employee. The automobile had recently caught his attention, and Dietrich reasoned that working for Brewster was one way to get involved in the burgeoning industry and the Mechanics Institute instruction would pave the way. To this end Dietrich spent twelve hours at the Mechanics Institute campus every week and received his diploma in 1918.

In time, Thomas L. Hubbard joined the Brewster roster of automobile designers. The two talented men, who were about the same age, became friends. During lunch one day, they discussed the possibility of organizing a company that designed and engineered automobile bodies. After leaving Brewster, the two partners established a clever new name, LeBaron, Carrossiers. Their business took off, and in time, they developed innovations in automobile design with an unmistakable sporting style. This firm would evolve into the quintessential automobile designers and producers in America and achieved considerable recognition worldwide.

Charles Henry Thornton Sr. entered the construction field as an electrician's apprentice at the age of fifteen. It was natural for Thornton to enter this field since his family had been involved with the construction business since 1846, when his father started a masonry contracting company.

A cousin, Peter A. Thornton, who had attended the Mechanics Institute (1913–16), urged Charles to attend the Institute. Charles graduated from the Mechanics Institute in 1927, the year following the completion of his bricklayer's apprenticeship. It is interesting to note that Charles H. Thornton Sr. and Anthony DeBenedittis graduated in the same year; both men with a specialty in architectural drafting.

Charles H. Thornton Sr. was initiated as a GSMT member in 1961 and is listed as a construction inspector. He served with distinction throughout his tenure as second vice president (1966), vice president (1967) and president (1968). His sons are also members of the GSMT: Charles H. Jr. (1963), William A. (1965) and Robert F. (1972).

At his inaugural address in 1968, President Charles H. Thornton Sr. said:

> *The General Society of Mechanics and Tradesmen of the City of New York is now in the 183rd year of its existence, the second oldest Society in the United States having been founded in the year 1785. Over the years our Society has given thousands of technically minded young men the opportunity to advance in their chosen fields. Among the list of former students are many who have brought credit upon themselves in government, industry and other stations in life.*
>
> *Mechanics Institute is opened to the Doers of today and the Achievers of tomorrow. Many of our nation's achievements of tomorrow are going to be born in the skilled hands and know how of our young apprentices of today. Mechanics Institute will find ever increasing pressure upon it to fill the need for trained skilled tradesmen. As in the past, our Society will do all in its power to meet the challenge.*

The capstone in the Thornton family's involvement in the construction industry comes full circle with Society member (1963) Charles H. Thornton Jr., PhD, PE, who is chairman of several organizations: Charles H. Thornton and Company, LLC; STRAAM Corporation; TTG Corporation; AECOS/TopDown; and Founding Principal and chairman of Thornton Tomasetti. Thornton is also co-chairman of ACE Mentor Program of America. The year 2015 marks forty-seven years of Thornton's involvement in the design and construction of millions of dollars' worth of projects in the United States and overseas.

MECHANICS INSTITUTE TODAY

"Build Yourself. Build New York" is the slogan of the tuition-free Mechanics Institute. Today, professionals with sophisticated knowledge of industry practices and state-of-the art technical innovations teach the classes. These instructors work in the major building trades by day and at night teach skills to enable students to advance in the building and construction industry.

Each year, students enroll in the Mechanics Institute's programs, including Virtual Design and Construction, Electrical Technology, HVAC Systems Design, Plumbing Design, Construction Project Management, Facilities Management and Historic Preservation. In step with career development, the Mechanics Institute offers the advantage of new technologies such as computer-aided drafting (CAD) and building information modeling (BIM).

As a result of this state-of-the-art training, the students maintain a competitive edge in a technologically advancing environment. The leading building and construction companies value the Mechanics Institute's students and graduates who continue to "Build New York."

Graduation 2014

Commencement exercises celebrating the motivations, efforts and accomplishments of the Mechanics Institute's graduates are held within the traditional and historic setting of The General Society's Library, located on the ground floor of the GSMT building.

At the Mechanics Institute's 2014 graduation, students proudly marched into the GSMT library for a traditional cap-and-gown ceremony accompanied by the requisite "Pomp and Circumstance" music. Family and friends joined the graduates to celebrate their accomplishments: completing two- and three-year programs at the Mechanics Institute.

The assembled students and guests were greeted by The General Society president and Society member (2008), Gottfried Weissgerber. President

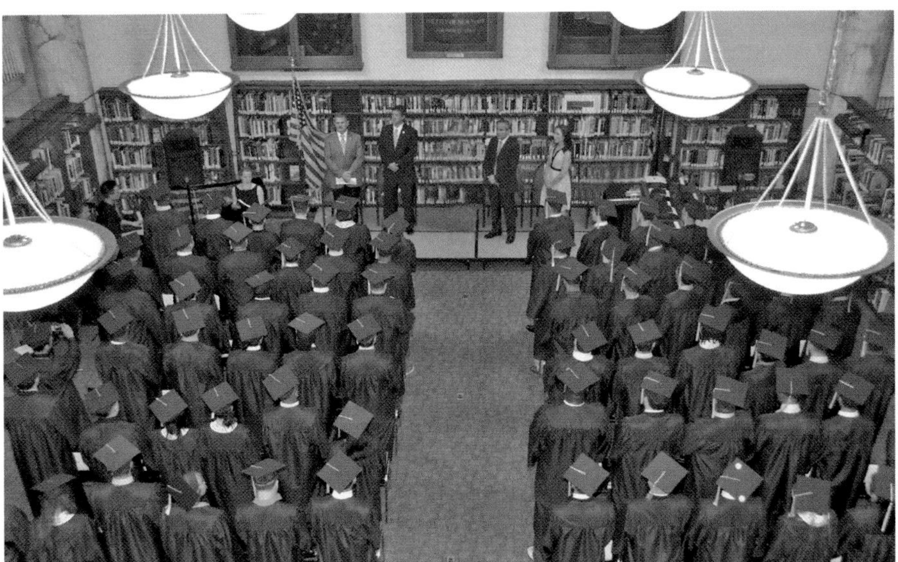

The Mechanics Institute's graduation celebration held in the historic GSMT Library. *Photograph by D.A. Luhmann.*

Weissgerber is a 1980 Mechanics Institute graduate and a George E. Hoe– award recipient. Weissgerber encouraged the graduates:

> *We wish you, our Graduates, great success in all your endeavors. You join a long legacy of graduates going back to 1858. As Mechanics Institute alumni we will continue to welcome you here at The General Society and hope that as you progress in your chosen profession, you will join with your fellow alumni in encouraging and supporting others toward this journey of accomplishment.*

The "Pledge of Allegiance" opened the ceremonies followed with the vibrant singing of the "Star-Spangled Banner." Victoria A. Dengel, executive director, introduced the keynote speaker, Louis J. Coletti, president and CEO of Building Trades Employers' Association (BTEA) who acknowledged the students:

> *You have done it the hard way going to work during the day and studying at night. Bring that same hard work to wherever you are designing or working on a project. You worked hard because you love what you do. Then take personal pride when you can say, "You see that building, I built that!" You are going to be building the future of this city. Work hard, the construction industry is changing hour by the hour. Technology is moving so quickly you have to keep your learning on trend. This is an industry where you will learn every day. Now you are starting to learn to be outstanding in your field. There is no better work force in the world than the New York City work force. The industry is looking for people. There's work out there. Your input will influence generations to come. Be proud of what you do.*

Speaker Coletti further charged the graduates with walking down our city's streets and proudly viewing the beautiful buildings, schools and institutions that had been built here by workers in the construction industry: "The New York City skyline is among the most beautiful in the world," he added, "and great skill was needed to achieve it. Few other industries provide their workers with such visible evidence of their accomplishments."

Flora Ramos—class of 2007 and currently director of community relations at Tishman Construction Corporation, an AECOM company—gave the alumni address, telling the students:

> *Congratulations! I am honored to be here. Construction isn't a field you expect to go into. Prior to working for Tishman Construction Corporation I worked for Pan American Airlines and during that time I became more interested in the Tishman Corporation. I applied my studies at the Mechanics Institute in Construction Project Management and you, the graduates of today have the same goals. Always give your best!*

The teachers at the Mechanics Institute also got their due praise. The valedictorian address by Ramon Encarnacion, graduating in construction project management, congratulated the students:

> *We finally made it! Today is a celebration of perseverance. We fought the good fight, we won! The key to Mechanics Institute are the professors. They have knowledge and bring us correct solutions. The teachers became mentors and motivators. The education provided by the Mechanics Institute gave us the latest technology, by the most current professionals in the marketplace.*

The presentation of medals and awards for outstanding scholarship included Alec Strasser, who received the George E. Hoe Award for Construction Documents Design from Society member (2012) Andis Woodlief, chairman of the GSMT school committee. Society member (1987) Keith S. Olsen presented the Albert A. Cuneo Scholarship Award, established by Keith's grandfather, Albert A. Cuneo, Society member (1950). The award is based on a graduate's overall excellence in their work. Raisa Zherebkina, an HVAC systems graduate and a recipient of one of the Cuneo awards, remarked about her experience at the Mechanics Institute, enthusiastically saying, "It helped to advance my career in mechanical web engineering, and I am proud of my progress and pleased to accept the award."

Another presenter, Kristine DeNapoli, president of KND Licensed Electrical Contracting & Services Corporation presented graduate Alrid Reid with the Thomas S. DeNapoli Award in Electrical Technology in memory of her father-in-law, Thomas DeNapoli. Kristine DeNapoli commented, "The Mechanics Institute has had a tremendous impact on my life. We are happy to give back as well as consider it a responsibility to do this."

A Society member since 2012, Alan Senzer contributed the Alan Senzer Award, which was presented by one of The General Society's past presidents Gerard Drohan Jr., who was reelected president in 2015. The award went to graduate Ramon Encarnacion, who studied construction project management.

After the awarding of diplomas, the assembled audience rose to sing "America, the Beautiful." Following the recessional, a fine buffet dinner was served, and the celebration continued with karaoke entertainment by Mechanics Institute graduate Andrew Bennett.

It is remarkable to know that, over the years, thousands of students have graduated from the Mechanics Institute. In 1931, at the height of country's Great Depression, enrollment in the tuition-free Mechanics Institute was at an all-time high. At the GSMT's 146th Anniversary celebration, Society member (1921) Edward Lawrence Hoffman, superintendent of The Society for forty years, greeted the guests. He talked about an exhibition of the work of the students then enrolled in the Mechanics Institute, "There may be an Andrew Carnegie or Henry Ford among these boys exhibiting tonight," and the same may be said of the 2014 graduates.

BUILD YOURSELF. BUILD NEW YORK

The Mechanics Institute spans over 147 years of continuous operation since it opened in 1858 as a tuition-free technical school, an educational program of The General Society of Mechanics and Tradesmen. It has provided evening instruction in trades-related education that improves job opportunities for New Yorkers in the building and construction trades. It is for this reason that the burgeoning construction industry values the Mechanics Institute graduates whose education in advanced technologies serves to build our magnificent city. A valedictorian speaker at a previous graduation summed his education up as follows:

> *Classes at MI had not taught them "what to think," but rather "how to think," how to ask questions, how to reject assumptions, and how to exercise reason in approaching a problem—valuable lessons in any field of study.*

Today, many graduates hold key jobs in the construction and building industry or have opened their own companies—these are testaments to the skills learned at the Mechanics Institute that have provided professional development for thousands of students who have adhered to the mantra, "Build Yourself. Build New York."

9

The Apprentices' Library and The General Society Library

A library, judiciously selected is the Garden of Eden, in which is planted the tree of useful knowledge, the fruit of which is life, not death. We have that library here, we planted that little branch, which has grown into a goodly tree, fair to the eye, and we must water it by our liberality until it bears gold fruit to the taste.
—*Mordecai M. Noah, 1850*

The General Society's Apprentices' Library, founded in 1820, had a parallel growth to the Mechanics School. Every time the school relocated, the library moved along to the larger quarters. The library was primarily established for the young men who apprenticed with master craftsmen, so the men had an opportunity to improve themselves through reading. Other reason for establishing the library, the Brothers said, "It was hoped that members, especially the apprentices, would spend more time reading books instead of spending time in bar rooms, theaters, and perhaps in bad company."

The esteemed members of The Society had also decided that a library was a necessity, since libraries that existed at that time in New York City would not allow the students of The General Society School to use their facilities. Furthermore, even if these libraries allowed students to enter their hallowed halls, they would not have found books that might interest them, particularly those dealing with mechanical and industrial subjects.

Although New York City had a lively press that turned out newspapers, pamphlets and broadsides, few printers published books, and most of the books available in New York were imported from England. In general, books were expensive and scarce, and only the wealthy could afford such luxury.

The Apprentices' Library had the same goals and objectives as the Mechanics School, and many of the same Brothers who were responsible for the founding of the Mechanics School had a hand in establishing the library. At a General Society meeting on March 1, 1820:

> *It was revealed that a Library Committee, consisting of Thomas R. Mercein be appointed to establish a library for the use of apprentices or mechanics of this city; and the committee was also empowered to issue an appropriate address to the inhabitants of the city, stating the object, and solicitation of funds.*

Thomas R. Mercein is acknowledged as the principal force behind establishing the Apprentices' Library. He was affiliated with The Society on several levels of executive management—member of The Society (1803), second vice president (1814), vice president (1815) and president (1827).

The library committee of loyal supporters also included such inspired men as Society member (1793) Francis Cooper, who was vice president (1802) and president (1805). Society member (1821) Mordecai M. Noah, also figured importantly in this group of visionaries who championed the library's usefulness.

Society member (1794) John Slidell Jr., who was vice president (1803) and president (1804), also joined the ranks of organizers for the library. However, The Society had other supporters, and many

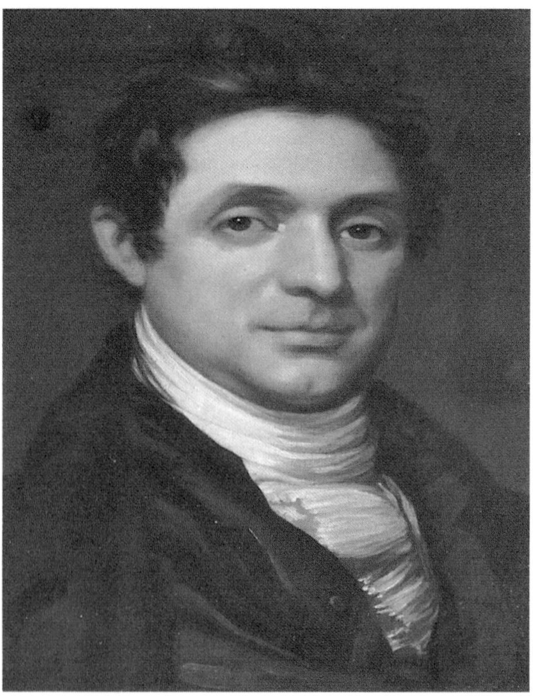

Society member (1803) Thomas R. Mercein, Apprentices' Library founder, vice president (1815) and president (1827). His portrait is located in the GSMT Assembly Room.

private citizens also endorsed the library project, such as John Pintard and William Wood.

The founding of the Apprentices' Library, as it was called in those early days, was not without controversy. Not everyone agreed that the apprentices would want to read or that this was the best use of The Society's money. The reason for the negative reaction was attributed to a few of The Society's members. They believed that the library would be a failure because apprentices and young people were more interested in spending their free time in taverns than in intellectual pursuits. Little did they realize how much the apprentices' thirst for knowledge would be quenched with those books.

Mercein's and Wood's Support

Despite naysayers, Thomas Mercein, chairman of the library committee, zealously worked on the project. On March 4, 1820, he wrote a letter to William Wood requesting advice. Wood was a merchant and philanthropist from Canandaigua, New York, who had played a leading role in establishing apprentices' libraries in Boston and New Orleans in the United States and Montreal in Canada. An excerpt from the letter reads:

> *Although, I have not the honor of a personal acquaintance with you, I am emboldened to address some inquiries to you in relation to establishment of the library for apprentices, lately established in Boston, because I am persuaded that the same enlarged and liberal views that induced you to take a leading part in the Boston Institution will prompt you to afford the benefit of your advice towards a similar one in this city*[.] *The Mechanics Society of this city (a wealthy and respectable Institution) have taken the thing up very spiritedly, and have appointed a committee (of which I am chairman) to digest and report a plan for carrying the project into immediate effect…Perhaps it would be more advisable, instead of making any inquiries of you, to beg that you will, in your own way, give us all the information of which you are possessed…a copy of any printed documents, blanks, etc., and if you see any thing to amend in the plan you so lately organized, please mention it…We wish to act promptly and efficiently, and…are persuaded (the library) will, if properly managed, result in effecting as great an amount of good as an institution ever instituted in this city.*

Wood subsequently traveled to New York and worked with Mercein to obtain the facilities and books for the library. It was Mercein's campaigning that resulted in The Society voting favorably for the project. In addition, Wood, a prolific supporter of The General Society's library, spent four months on the campaign trail and wrote numerous newspaper articles in an attempt to attract public support for the library. To this end, Wood went on a tour of more than one thousand mechanic's shops and small industries for membership in the library. He also solicited the social elite and influential people for contributions of funds and books. To establish its library, The General Society also solicited support and donations of books from the mechanics of the city.

In 1820, the library committee reported:

> *That of the whole male population of this city, at least 6,000 are mechanics, most of whom have from one to four apprentices, who are generally the children of those parents whose situation in life have rendered them unable (however willing) to afford that education and instruction required to render them useful and efficient members of society; hence every means that can be devised to further the improvement and advance the knowledge of such, ought eagerly to be seized, in order to remedy, so far as practicable, the privation suffered in their early years, and which, in the opinion of your committee, cannot be more effectively promoted than by affording to all who are desirous of improving themselves, the means by which to attain that object, to wit: the gratuitous reading of elementary, more, religious and miscellaneous books, and such others as may have a tendency to promote them in their several avocations.*

At a subsequent meeting on January 11, 1820, another plan for the creation and maintenance of a library was received and adopted.

THE APPRENTICES' LIBRARY

The library was officially opened on November 25, 1820, on Evacuation Day, an important patriotic holiday that was celebrated for years to commemorate the British withdrawal from the city after the American Revolution. Though the weather was inclement, New Yorkers celebrated the patriotic holiday in tandem with the opening of the Apprentices' Library, and a large audience attended the ceremonies, including the mayor and members of the common council and state legislature. As befitting such an auspicious occasion, author Tom Glynn wrote the following excerpt of Thomas Mercein's keynote address, stressing:

> *The importance and respectability of the city's artisans and referred to the library as one of "the securities which we are planting around the fortress of Liberty, erected in the glorious and triumphant struggle of the Revolution." The library would be a place where young men could acquire "those sound and commendable habits, that will mold the character, and elevate it to a standing, equally congenial to individual and general happiness." However, he also emphasized other values more closely associated with the liberalism of the emergent middle class. Mercein assured the apprentices in the audience that their "opportunities (were) great and liberal;" that "industry, ardour, sobriety, and perseverance will lead to successful competition… (and) prosperity." Mercein urged the apprentices to "avoid the alluring but fatal paths of vice and dissipation: and always to be sober and industrious.*

Mercein implied a practical, vocational purpose for the library, calling it a "source of rational and useful information," and extolling "new combinations and new discoveries constantly developed in the useful arts. It was also thought that with the advantages of a free Library, a young apprentice could compete successfully in the marketplace and thus contribute to the general prosperity of the city."

The library went on supplying books to the working young men of New York. The shelves contained a cache of books, and the tables were covered with newspapers and periodicals, including early eighteenth-century periodicals such as the *London Magazine and Monthly Chronologer* (1736). On the opening of library in 1820, Society member (1832) Arnold Douglass recorded his name as a reader and took home Bunyan's *Holy War*.

The library began in rooms rented from the New York Free School Society on Chatham Street. At its inauguration, the library consisted of three large boxes of books. After dark that evening, under the cloak of candlelight, apprentices whose employees had vouched for their character checked out nearly three hundred books. By the following Saturday, all fears that the apprentices would not avail themselves of the advantages of the library were dissipated.

Within two years, The General Society purchased its first school building at 10–14 Chatham Street, and the cornerstone was laid with appropriate ceremonies on June 13, 1821, the members of The Society, the readers of the library and children of the school participating. The plaque read:

> *This building erected by The General Society of Mechanics and Tradesmen of the City of New York, for a Mechanics School and Apprentices' Library, A.D. 1821, and in the 45^h year of the Independence of the United States of America.*

By 1822, the Apprentice Library had six thousand volumes and 1,300 members. Nearly 1,000 of those readers were male apprentices between the ages of twelve and seventeen.

Rules and regulations adopted by the Apprentices' Library were strictly enforced. Borrowing privileges were available at no cost to any apprentice who was able to "produce a certificate, signed by his employer, which stated that he was employed, that he was allowed to use the library, and that his employer would vouch for his character and guaranteeing the safe return, in good order of all books." As the collection grew, the books were kept in locked cases. The librarian, as custodian of books, enforced the rules. According to the 1822 minutes of the library committee, rules were set as follows:

> *Applicants must come with clean hands and faces and on entering the room must be seated until called up. No bad language, loud talking, or smoking allowed. Each boy must leave the room on receiving his book and must not play or remain around the door.*

Strict rules of conduct went in tandem with the loan of books, which circulated for two weeks and could be renewed. Overdue fines were assessed at three to twelve and a half cents a week, depending on the size of the book. The borrower was liable for the entire value of a book not returned within a month. There was also a fine for damaged and mutilated books. The same rooms that were used for the Mechanics School were also used for the library, and each time the Mechanics School expanded, the library was given larger quarters.

An early library committee report declared that over eight hundred apprentices were using the library and that application for books became so numerous as to require the library to be open every day of the week. To accommodate the apprentices' work schedule, the library was open from six until nine in the evening.

According to a newspaper account in the *Evening Post* on November 27, 1820, the greatest demand of books included the chronicles about Benjamin Franklin and George Washington, the history of the United States and the War of 1812. Other subjects in demand were histories of Greece, Rome and England and voyages. The library also purchased technical books on architecture and mathematics, but most of the books were donated by members and wealthy New Yorkers.

Committee reports documented the progress of the library. For instance, in March 1823, the membership learned that there were two hundred constant readers among the one thousand apprentices regularly receiving books. There were six thousand volumes available, many of which needed to be rebound and replaced.

GROWTH OF THE LIBRARY

The City of New York also encouraged the growth of the library. On February 16, 1824, The Society asked The General Council of New York City:

> *To relinquish one-hundred and twenty-five dollars which was due for the annual rent paid to the city for the land on which a newer projected library was to stand, for the reason that The Society was unable financially to extend the library without the extra funds. And on December 6, 1824 a resolution of thanks was received by the city from The Society for the generous relinquishment of five years of ground rent on which their institution was erected. This aid was considered a contribution, in aid of funds for the purchase of books for the use of the library, and The Society informed the city on December 19, 1825 that it had purchased with the money donated by the city two hundred and forty-two volumes.*

A catalogue published in 1835 showed that works in mechanics, chemistry, mathematics, natural philosophy and political economy dominated the library collection, although it also included some works of biography, history and fiction. The reading room offered current newspapers and over fifty periodicals in English and a few other languages, along with nine daily newspapers.

The reading room was a popular rendezvous place, and the directors reported that it was "almost constantly occupied by numerous readers." In 1841, Messrs. Bartlett and Wellford donated seventy-seven volumes of the

London Monthly Review. The library committee was constantly searching for ways to improve the library's usefulness, so in the same year, the library committee recommended that apprentices and journeymen should not be deprived of use of the library since "this is a period of life when they are in even great danger of going astray" and are "no longer subject to restraints and supervision." New York State also helped the library:

> *An act passed on February 7, 1842 by the State Legislature enabled The Society to use any profits derived from its school for the support and improvement of the library and reading room. The Society was also authorized to establish and hold a separate fund for the support of an Apprentices' Library and Reading-room, and to receive and hold any donation, devise, or bequest for the purpose of such fund, which shall be held and applied by The Society as in trust for these special purposes, and not be applied or diverted to other objects.*

On January 3, 1844, William Wood donated five hundred copies of the New Testament to be distributed among the readers of the library. A short time after, Wood donated a sum of money in relation to which he writes:

> *I now send on the Mechanics' Exchange Bank for $38.25 to buy, at The Society Bible House 475 Testaments same as the last, which will amount to $33.25. The five dollar over I beg, sir, you will do me the favor to deposit in the savings-bank, for the reader at your institution who may gain it under the following plan (to induce them to read a little more than they otherwise should). Every apprentice who has read the Testament throughout by the 15th of May next, to draw a lot in the library for the $5.00 which is offered to the fortunate one as a premium for his industry, etc. in acquiring knowledge from the "Book of Life."*

Eventually, the Chambers Street building was too small to accommodate both the school and library, and in 1837, with The Society's move to Crosby Street, a wing was added to each side of the school building and courses of lectures were established for the winter months. Entrance accessibility was at the core of The Society's subsequent decision. The Society presently desired an entrance to its library and lecture room from Broadway, and in 1845, the building at 472 Broadway was altered to connect in the rear with the Crosby Street property. The Broadway building's façade was duly ornamented with the Hammer and Hand insignia.

The Society's policy of renting space in their buildings was a sound decision, albeit one that met with some criticism. At this time, an unexpected source of income came through leasing the lecture room to the original Christy Minstrels when it was not in use. Numerous members found this entertainment offensive and regarded this scheme with disfavor, yet this rental fueled the coffers of The Society.

The establishment of the Apprentices' Library and the school was part of a new burst of interest at the time to use education and reading as tools to prevent licentiousness. Many master craftsmen began challenging the "antisocial evils" of their workers, such as swearing, drinking, Sabbath-breaking and gambling. Just as the Apprentices' Library had established strict codes of conduct and deportment, other institutions began to require strict work codes that demanded a new bourgeois standard of order and personal decorum. Merchants and entrepreneurs, including members of The General Society, targeted the laboring population to teach good habits and skills.

Abram Stevens Hewitt's Visit

One apprentice of note who had attended classes at the Mechanics Institute was Abram Stevens Hewitt. A diligent scholar, Abram spent his time on downtown Crosby Street at the Apprentices' Library.

At the 1885 Centennial Celebration of The General Society of Mechanics and Tradesmen, Abram recalled a visit to the library with his father, Society Member (1810) John Hewitt, the cabinetmaker. Abram, as mentioned in a previous chapter, was a member of The Society in 1888 at the height of his career success, when he was mayor of New York City. Here is Hewitt's recollection of that day in 1833 when he first visited the library:

> *When I was a boy of eleven years of age, The Society removed its library to Crosby Street my father took me by the hand and led me up into the Apprentice Library. I was captivated; for the first time in my life, I saw books beyond the wildest dream of my fancy. I said to my father, "I am your apprentice, so that I can have some of these books?" He replied, "You can take books; you can take them on my account." Then and there with the opening of the library in Crosby Street I took my first book; I took it home and I shall never forget the impression that that book made upon me. I had*

> *heard a great deal about Shakespeare. I have been told that Shakespeare was the great poet of the English language. I wished to begin with him, and I took my first volume home with me. The first play that I read was "The Tempest." I diligently took volume after volume, until I had read every play of Shakespeare. I am bound to say that while, at the age, I could not understand fully the scope of what I read, that nevertheless the books that I thus took from the Apprentices' Library developed a taste for study which was the foundation of other tastes, and, until I went to college I was a regular participant in the advantages of the Apprentices' library.*

During Abram's early years, despite his narrow circumstances, he had the innate advantage of a steadfast character and brilliant mind. At the Apprentices' Library, he indulged his passion for reading in one of the richest collections of books in New York. As a result of his voracious reading habit, Abram had educated himself on several levels, which prepared him to attain two scholarships to attend Kings College (now Columbia University).

Inviting Female Readers

The library committee continued to search for ways to improve the library's usefulness. In 1841, it was resolved that journeymen should not be deprived of use of the library since "this is a period of life when they are in even greater danger of going astray," and in the 1847 resolution, the library and reading rooms of The Society were extended to educate deaf mutes. In a gesture that bears witness to The Society's advanced thinking, the committee took into consideration that the great storehouse of books in the library should be used to further advantage, particularly for females. The chairman reported:

> *That the Apprentices' Library, with its reading room is in good order and successful operation: however, that the number of its readers should be increased, for we all know that the influence of the library should be felt in every workshop in the city. And here, allow me to call attention to a class of operatives for whom no provision of this kind has ever been made—and here I allude to the large number of females engaged in the various employments connected with the mechanic and manufacturing interests of our city. We can, at very little expense, provide the means of intellectual improvement to them, and it is very desirable that we should do so. I am satisfied that it is*

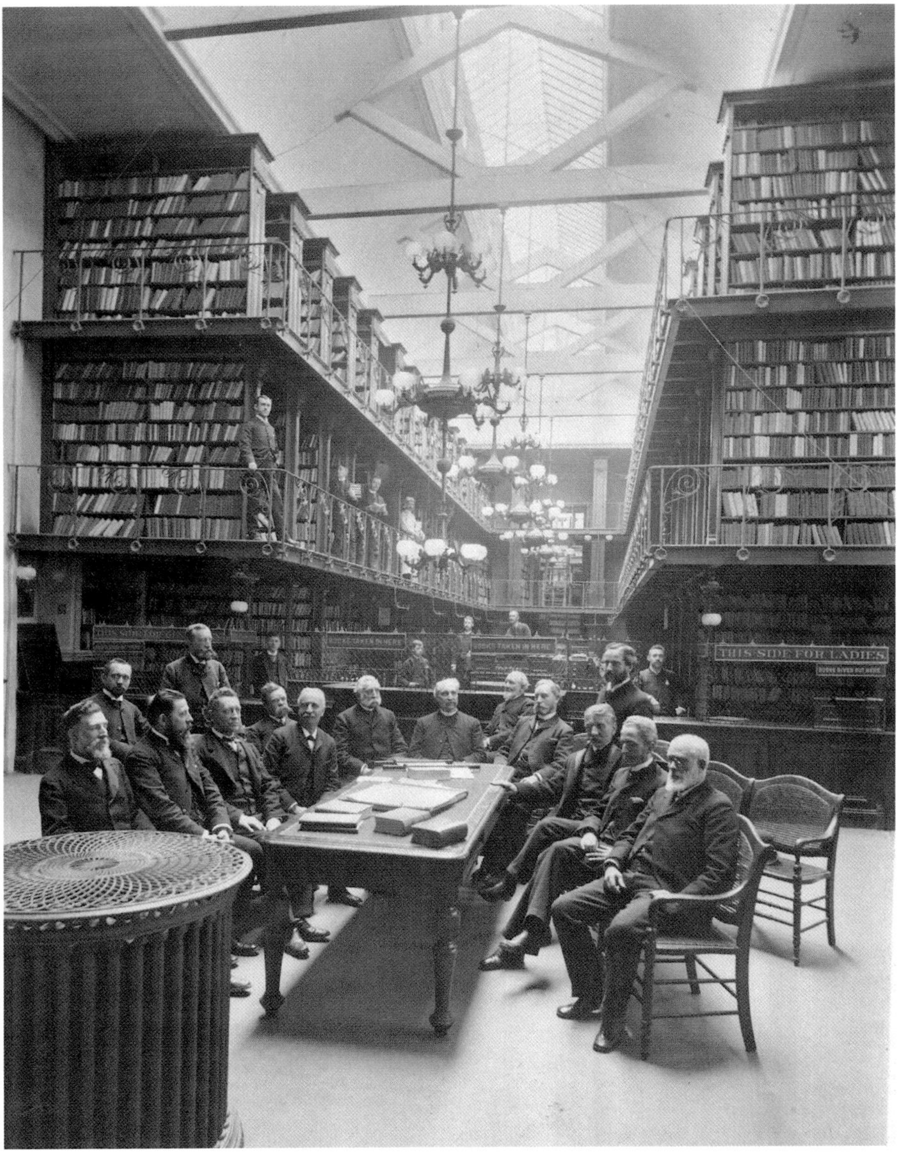

Library committee in the GSMT's library located in its fourth location at 18 East Sixteenth Street, 1878.

clearly within the legitimate objects of our association, and do not hesitate to recommend that our library be at once thrown open to female operatives, upon the same condition as it now is to apprentices.

To strengthen this point of view, the chairman gave the following example to encourage including women as library members:

> *I hold in my hand the Report of the Apprentices' Library Company of Philadelphia, for the year ending March 13th, 1855. I suppose that there was one published since, but I have not seen it. Both boys and girls have access to that library, and the report states that the number of volumes is about 15,700. The report also states the number of this interesting class of our readers continues to increase. It appears to me that any argument that can be advanced in favor of providing the means of intellectual improvement to boys applies with equal force to the same provision to girls.*

This wise counsel led to the Apprentices' Library's doors being opened to female apprentices, who were given library privileges in 1862 on the same basis as male apprentices.

The Society also realized that the library building at 472 Broadway was too far downtown to draw a large number of readers. Setting their sights on drawing in more library attendees, the Apprentices' Library moved to yet another site in 1877, the Suydam mansion at 18 East Sixteenth Street.

THE LIBRARY AWARD

The General Society continued to be at the forefront of representing itself at major patriotic, national and international events, such as the 1893 Columbian Exposition in Chicago that commemorated the 400th anniversary of Columbus's discovery of America. Chicago became known as the "White City" and the white, neoclassical architecture of the exhibition buildings influenced the design of public buildings for several decades. It was here that the eminent sculptors August St. Gaudens and Frederick William MacMonnies and the architects Stanford White and Louis Sullivan won recognition. The electrical display ushered in the age of electricity, and the value of structural steel was demonstrated. Many other exhibits presented innovations in production, manufacturing and administration.

According to the GSMT annals, The General Society's exhibition at the World's Columbia Exposition in Chicago was "so marked by character to attract universal attention and elicited an unsolicited award for a large career of usefulness, excellent organization and appliances and instructive and useful display."

> *During the year the library was visited by several persons connected with other libraries, some in foreign countries, whose attention was attracted by the exhibit our society made at the Worlds Columbian Exhibition in Chicago, in 1893, while the fact that our library was awarded a medal for the best Charging System, created a desire on their part to visit this library and to inspect, as they did, our system of charging and classification of books, etc., with which they expressed great satisfaction, and thanks for the courtesy shown them.*

Benevolent Donors

As time went on, several benefactors donated books and funds to improve the library's collection. For instance, Society member (1826) Benjamin DeMilt—a watchmaker, vice president (1831) and president (1832)—held a stake in the extension of the library's holdings. After his death in 1845, it was revealed that he had bequeathed his large personal library, adding 1,833 volumes to The Society's library. The library designated this collection as the DeMilt Section devoted to valuable books for reference. Shortly thereafter, DeMilt's daughters made very substantial bequests to The Society. Elizabeth DeMilt donated $10,000 and Sara DeMilt $2,000, making possible extensive alterations to the library and the Hall. Looking for ways to increase the usefulness of the Apprentices' Library early on, Benjamin DeMilt had suggested displaying miniature specimens depicting each apprentice's pursuit together with collections of minerals and foreign coins.

Peter Lorillard, an industrialist and real estate tycoon, owned tracts of land on which Tuxedo Park, the elite private enclave, was developed. He was initiated into The General Society in 1811. He, too, endorsed the library. Later, in 1868, $5,000 was received from his estate to be used for the purchase of books.

The holdings of library were greatly increased on subjects of shipbuilding. Society member (1841) William H. Webb—founder of Webb Academy and Home for Shipbuilders in Glen Cove, Long Island, New York (now the Webb Institute)—donated one thousand books in 1874.

Other donors were finding the library an important repository for special collections. In 1888, Laura H. Slade donated the J. Morgan Slade collection of architectural works, a collection of 128 volumes, together with photographs and prints. The Slade Architectural Section, devoted to rare and useful works on architecture and related subjects, is a rich depository of reference materials. The Slade Architectural Section, like the DeMilt Section, offered opportunities for apprentices to broaden their knowledge with deeper study.

This pattern of donating continued with contributions by John Jacob Astor and other titans of industry. The Library, as mentioned earlier, was also enriched by Andrew Carnegie's gift of $250,000, which made possible a reconstruction of The General Society building wherein extensions were added of the building.

Throughout the years, numerous smaller contributions were made to the library as well as donations for the purchase of books.

With the establishment of the New York Public Library on Fifth Avenue, however, The Society realized that the Apprentices' Library would be overshadowed in an effort to supply the public with popular literature, and that The Society should devote its efforts to supplying books in special lines, kindred to the Mechanics Institute's studies and demands.

THE LIBRARY'S PROGRESS

As recorded in the annals of The General Society, the library was able to exhibit phenomenal growth from the first 400 volumes in 1820 to some 14,000 in 1830 and 100,000 in 1900. The GSMT's library had an important impact on the formation of other mechanics' libraries and free circulating libraries throughout the United States and England.

The annual financial statement of 1872 showed properties were leased satisfactorily and other receipts generous enough to relieve any stringency of obligations incurred by The Society. The land at Broadway and Park Place had been leased long term, and a substantial building had been erected. The properties on Chambers Street and at Broadway and Crosby Street were in good condition.

Charity and taking care of its members continued to be at the core of The Society's mission. At this time, several burial plots were purchased in Cypress Hill Cemetery to be made available for those members or their wives who might die with no provision having been made for their last resting place.

The most agreeable part of the committee's work, however, consisted of receiving the many expressions of gratitude from the recipients of its bounty, some of which are quoted here. A daughter wrote:

> *It is indeed gratifying to us to know that you have set a stone at my father's grave, and I assure, both for my mother and myself, that we are deeply grateful for all you have done for us and to honor my dear father's memory.*

A son wrote:

> *Mother was for many years, a pensioner, on your society, and if you but knew how much comfort and assistance your benefits were to her you would rejoice in the good work you have done.*

A member of The Society, Brother B, wrote:

> *Yours received, enclosing check, for which accept my grateful thanks for the kindness you have shown and consideration given my case.*

The Mechanics School and the Apprentices' Library had an impact on the growth of similar institutions elsewhere. George Birkbeck, the founder of the Mechanics' Societies of Great Britain, stated that the great success and usefulness of the Mechanics School and Apprentices' Library in New York City prompted a decision to establish a mechanics' and apprentices' library in Liverpool and a school in Glasgow.

The General Society Library

By the end of the nineteenth century, the Apprentices' Library was the nearest thing to a free public library the city had between 1886 and 1898; a portion of the book collection was subsidized by public funds.

In 1858, the name Apprentices' Library was abandoned, and the name Free Library of The General Society of Mechanics and Tradesmen of the City of New York took its place. Then the name was eventually abbreviated and renamed The General Society Library.

The historical collections contain rare books, some first editions, and periodicals in all subjects, especially the "useful arts." It is interesting to note

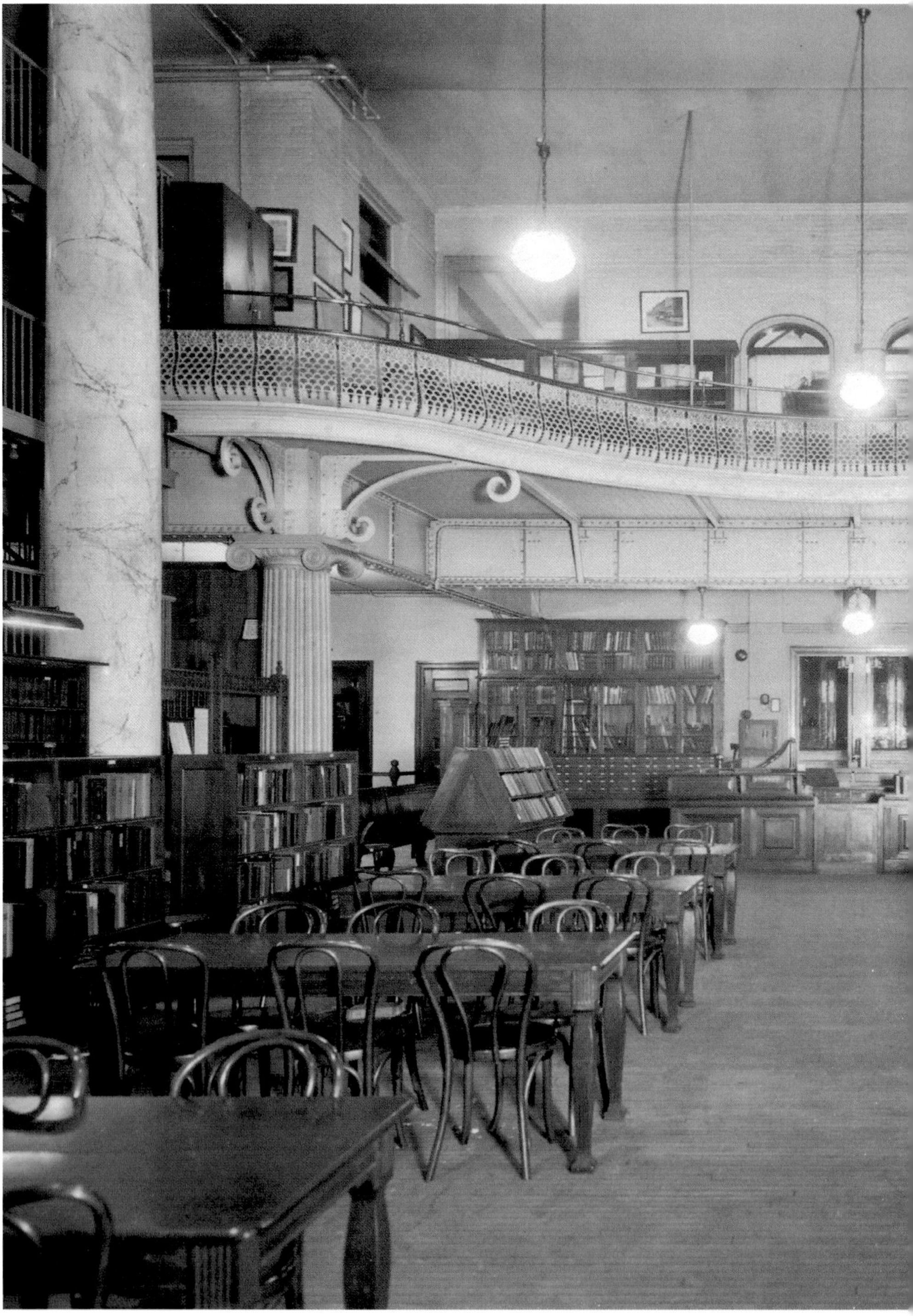

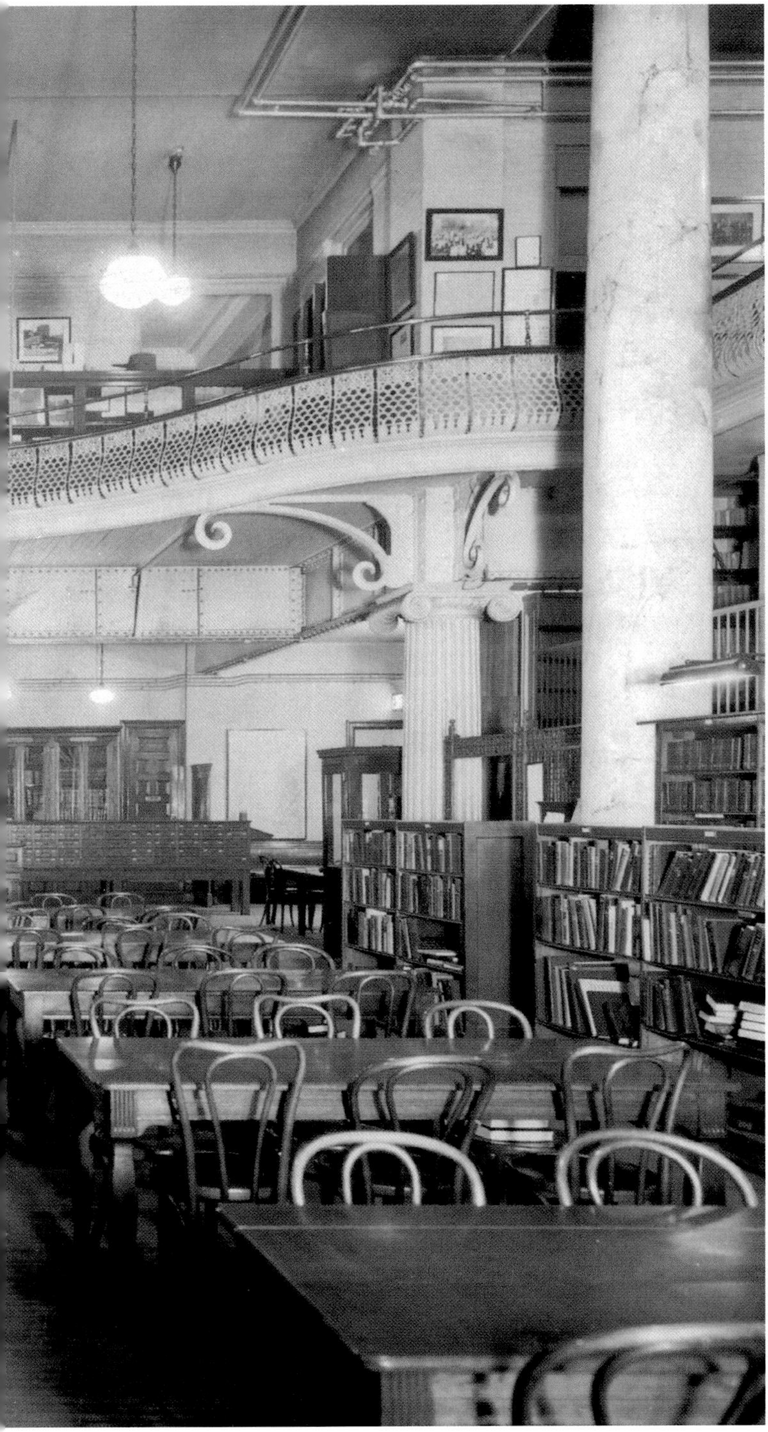

The Library at its present location, 20 West Forty-fourth Street, with its traditional library tables, 1936. Above the far balcony resides The Society's museum.

that the Library operated as the largest free circulating library in the city prior to the founding of the public library system and is one of the three remaining private-membership, circulating libraries in New York City.

The book and periodical collections of the Library span two centuries, number more than 100,000 volumes and are suited to both scholarly research and recreational reading. The library also makes its extensive collection available to other institutions and is open on a regular basis to the public who have become library members.

The non-circulating collection includes books and periodicals on the "useful and mechanical arts," urban trades and crafts, printing and publishing, building and construction and nineteenth-century America. The Library's mission is to:

> *SUPPORT the educational programs of The General Society; PRESERVE AND BUILD on the historic collections of The General Society; PROVIDE ACCESS to the collections of The General Society for researchers and the public; PROMOTE an understanding of urban work and its history.*

Recent acquisitions in the Library collection focus on the building environment and the occupations and activities of behind-the-scenes New York, including the building and construction industry, architecture, historic preservation, printing and publishing, labor and work history and cultural production.

The archives, one of the great treasures of The Society, contain the historical record of The General Society of Mechanics and Tradesmen of the City of New York, beginning with its founding in 1785, as well as other collections reflecting the work-history of the city. The archival collections, however, are open by appointment to qualified researchers with support from grants from the National Historic Publications and Records Commission and the National Endowment for the Humanities.

One of the most remarkable keepsakes in the archives is the historic GSMT's Registry Book, which records the names of every member who has joined The Society since its founding. Signatures since 1786 include The General Society's first chairman, Robert Boyd, and members Peter Cooper and his son-in-law Abram Stevens Hewitt. To date, there are over 4,400 signatures, attesting to ever increasing interest in The Society.

The Mechanics Institute collection includes important and unique material on art, architecture, drawing, design, drafting, architectural and mechanical drawing, jewelry design, printing, philosophy and lithography. The Luca Valentino Collection adds yet another storehouse of 350 books on cabinetmaking and furniture, and the Henry Hope Reed Collection is a gift of books from the personal library of the noted architectural historian. Other historic treasures reside in the circulating collections, including the Builder's Library, which focuses on the built environment, urban trades and crafts, labor and the history of the city. The list goes on, but I would be remiss if I did not mention the New York Civil War Roundtable Collection, in partnership with the New York Civil War Roundtable, which contains books on Civil War history.

The Library of The General Society, rich in history, chronicles the rise of the artisan class and their struggle and determination to become better educated and politically aware. More importantly, the Library's books and its archives are an infusion of history and heritage that underscores the importance of working people in the city's past.

GSMT's Permanent Home

The question of a new location for The Society was again raised. The inaugural address of Society member (1851) John T. Conover, president of The Society (1866), expressed The Society's concern:

> *Whenever and wherever you may build, I never want to see this Society build until they have made up their minds to erect a building that will be the pride of the mechanics, an ornament to the city, and monument to the mechanical genius and skill of the nineteenth century.*

The GSMT's fifth and final home at 20 West Forty-fourth Street upholds the standard set by Conover. Through the years, the organization has extended its reputation as a quintessential center of cultural activity. As you enter the GSMT building, which The Society acquired in 1899, your attention is immediately drawn to the building's landmark façade, which has been well preserved throughout the years. As mentioned earlier, it was designed by the architectural firm of Lamb & Rich and built in 1890 as the Berkeley School. Architect Ralph Samuel Townsend designed renovations to the building from 1903 to 1905. Today, the Library is updated with technology and offers Wi-Fi access to students, library members and researchers.

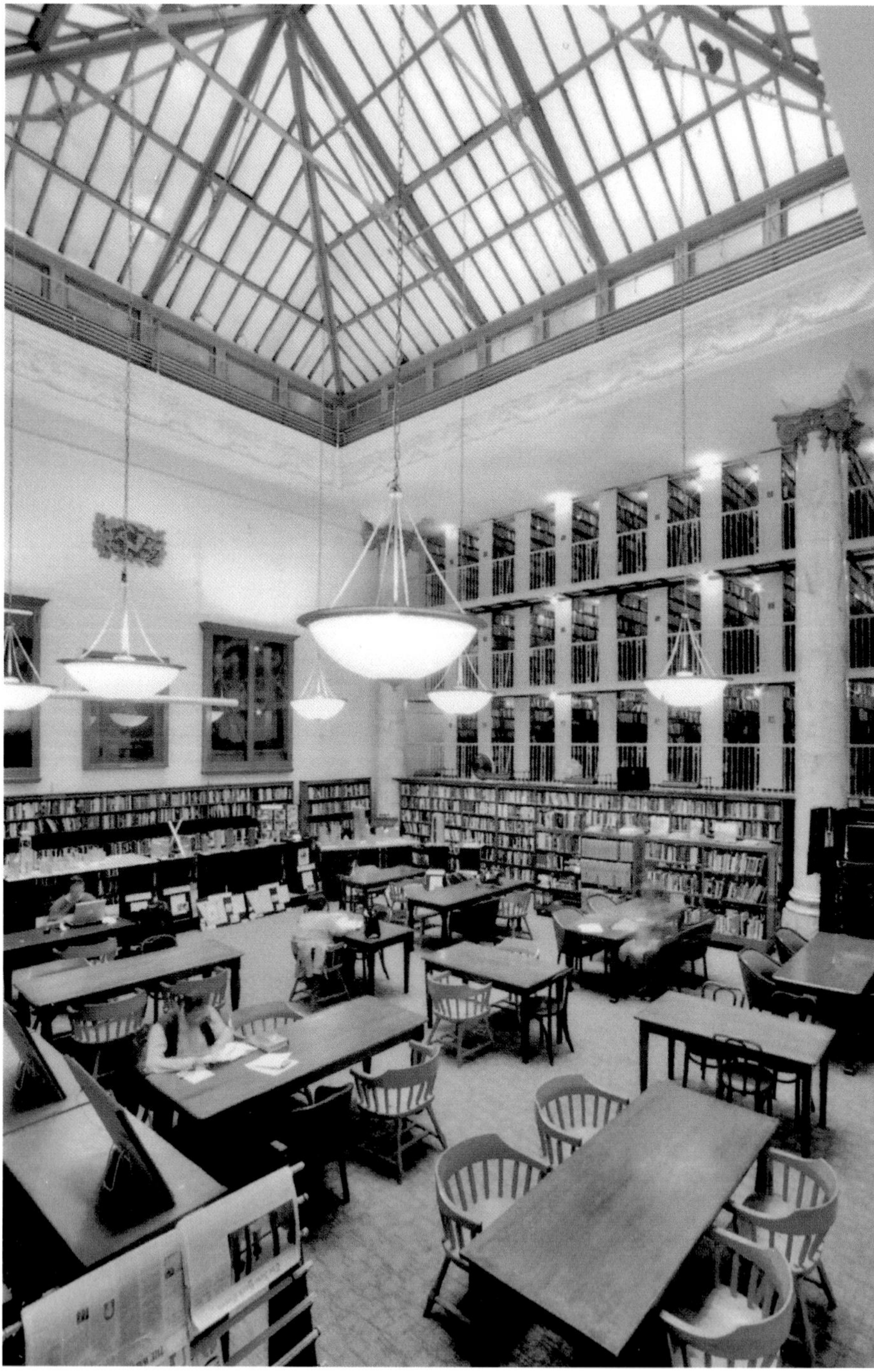

Through the GSMT's portal, thousands of young men and women have come and gone on to greater success through the benefit of books and schooling. In its present location, the GSMT supports its historical legacy with the educational programs of The Society, which include its school, The Mechanics Institute; the Labor, Literature and Landmark Lecture Series and the Artisan Lecture Series; and other innovative programs.

Currently, the Library has extended its reach far beyond earlier expectations and reinvents its original purpose by providing new exposure and functions as a site-specific venue for numerous cultural programs, television and film productions, designer fashion shows and concerts. Most notable, the Library is the destination location for patriotic and social celebrations that uphold the American tradition of commemorating Veterans Day, the Fourth of July and Flag Day.

Opposite: The General Society Library today has multi-level stacks, a soaring skylight and a Wi-Fi connection for Mechanics Institute students, library members and researchers. *Courtesy of Stephen Amiaga.*

10

The Lecture Series and The Small Press Center

I am convinced that the courses of lectures given by The Society have done much to educate the young men of this country in the application of science to business.
—Abram Stevens Hewitt

The General Society of Mechanics and Tradesmen, ever cognizant of the tenor of the times and with interest in developing cultural programs, considered offering a series of public lectures. As early as May 1, 1833, at a General Society meeting, a suggestion was made by one of the Brothers that, since The Society's charity needs were met and its school "had been awarded the prize of being one of the best schools in the State," attention might be drawn:

> *To that part of the community with whom we are immediately connected and to whose improvement and advancement we are most particularly pledged, that in addition to what this Society has already done, popular lectures upon subjects which we are interested in, as members of this society, and of the great Republican family of which the Mechanic and Tradesman form so important a part, and a convenient lecture room (should) be provided.*

In a report from the Committee for Promoting and Disseminating Literary and Scientific Knowledge, a plan to erect a separate building was recommended for a lecture room on the third floor of the Mechanics Hall to seat some seven hundred spectators. This plan led to the purchase of the land at 472 Broadway, also known as 30–36 Crosby Street, to be ready for the 1837 lecture season scheduled in the GSMT's newly built lecture room. The committee stated:

> *In this enlightened age, when the improvement of the mind and the diffusion of knowledge seem to embrace such a large sphere of action, and to occupy the attention of all classes of society, the propriety and usefulness of this mode of communicating instruction will hardly be questioned.*

Thomas R. Mercein, chairman of the Committee for Promoting and Disseminating Literary and Scientific Knowledge, congratulated the institute for being:

> *A place of meeting for business and for mental improvement alike, appropriate and worthy of the character of The General Society of Mechanics and Tradesmen.*

The Lyceum movement in America grew up around libraries as a "cultural adaptation" and "assumed the form of the lecture platform." At a time when self-improvement was the order of the day, The General Society's new lecture program met a social need by acquainting citizens with expanded knowledge of the natural and social sciences. The lectures became the height of social entertainments and galvanized the inquiring public with scientific and literary topics by leading scholars and professional orators.

The Lectures Inaugurated

The first lectures presented by literary and scientific committee included speakers culled from The Society's membership as well as notable professors and scholars. In a report from November 1, 1837, the committee believed:

> *That a full course of lectures on chemistry and particularly as applicable to the Mechanics' Arts, and other practicable purposes, would alike redound to the character of The Society and be acceptable to the members.*

The committee included lectures in fields of chemistry and science where significant changes were occurring. To this end, they made arrangements with Dr. John Torrey for a full course of fifteen lectures, embracing a general view of the most important facts and principles of chemistry and the practical applications of the science to the business of life.

Dr. Torrey's first lecture, "Chemistry, and the Application of Science to the Business of Life" was held on November 14, 1837, and the lecture series that began that month was to be heard every Monday and Thursday evening. Adding to viewer pleasure, it was reported that Dr. Torrey's apparatus was very complete and the accompanying illustrations were numerous and interesting.

In the same year, Professor James Renwick of Columbia College delivered ten lectures on natural philosophy; Reverend Dr. Gardiner Spring delivered a talk called "The Importance of Industrious Habits to Young Men;" and Society Member (1821) Mordecai M. Noah delivered the lecture "The Rise, Progress, and Downfall of the Moorish Race." So diverse were the subjects that Christopher Dunkin was engaged to deliver a talk called "Phrenology as Applicable in Education." From time to time, lighter subjects were also included, such as a Professor Barber's lecture "The Character and Writing of Lord Byron."

At a subsequent meeting, held on November 7, 1838, the literary and scientific committee seemed satisfied with its foray into the new lecture program and its outreach to the public, students, family and friends of The Society. The committee reported that arrangements had been made to continue a course of lectures on every Tuesday and Thursday evening of each week during the coming season. Abram Stevens Hewitt recalls:

> *I attended the first course of lectures given by Professor Renwick on chemistry. I am convinced that the courses of lectures given by this society had done much to educate the young men of this country in the application of science to business. The influence of the society was certainly very great, indeed, when it moved one of its members to found the Cooper Institute.*

By 1838, subjects varied widely, and several men engaged as the lecturers stepped up to the podium and delivered worthy messages on diverse subjects: Dr. John H. Griscom was engaged to deliver four lectures, "The Mechanism of the Human Frame," the subject of which was the delicate mechanism of the human frame, elucidated and made entirely intelligible by numerous drawings, models and preparations; Peter Harvey gave three lectures on various subjects, including one on moral philosophy. To cover the expenses, a charter revision was procured to allow ten dollars of each thirty-dollar membership initiation fee to be allotted to a special fund supporting the lecture series. Tickets were distributed to Society members for use by the entire family, and the privilege extended to teachers in the Mechanics School "both male and female…with the hope that the valuable information may be imparted to the students in their charge," Mercein declared.

These lectures were well attended and proved so acceptable that similar lectures were scheduled through the winter months.

Lectures and Education

Lectures were a popular diversion of the day, and many scholars and orators made a career on the lecture circuit. At the GSMT, the scheduled lectures became more than mere educational vehicles; they were highly coveted social events. It was an occasion for guests to show off their finest, and the social elite took advantage of the occasion. Dignitaries arriving in carriages were dressed in festive attire. Other guests, not so showy in their decorum, happily mingled with the elite in the audience.

The General Society, with its prestige and reputation, could cull from the best of the scholars and professors among its members and contacts in the city's cultural institutions. The general population, meanwhile, was eager to expand their knowledge, and self-improvement was a popular trend. As a result, The General Society lectures became increasingly popular, and a broader scope of subjects was considered.

By 1846, lectures were being given on architecture, Socrates and chemistry by notable men, pillars of society and professors. Speakers included the Honorable George Folsom, Esq. (member of the state senate), whose lecture was not listed; Professor George R. Gliddon spoke on "The Antiquities of Egypt"; Samuel Wells Williams, who lectured about China; Dr. Gunning S. Bedford gave a lecture called "Man Physically Considered"; and A.W. Armstrong, who orated on the topic "Self Education."

The lectures continued to attract guests in record numbers, chairman Thomas R. Mercein, on behalf of the literary committee, reported:

> *The lectures will commence on each evening at half past seven o'clock precisely, and the Committee respectfully, but urgently recommend to all who do attend, to be punctual to the time. It is extremely painful and embarrassing to the lecturer to be constantly interrupted, as was*

the case last season, and it is at least ungenerous for a few to inflict on the hundreds who are punctual, the pain and inconvenience resulting from their irregularity. The experiment of the last season also rendered it necessary to direct that children under ten years of age are not to be admitted, and that those who are admitted should be seated with their parents or friends, and not together.

The literary and scientific committee kept pace with the interest in the lecture series, and on January 5, 1850, the committee scheduled six lectures presented by Professor Louis Agassiz called "The Natural Relationship Between Animals." Dr. George Bethune spoke about Holland and Dr. Christopher Anticell on chemistry.

The list of exceptionally able and prominent men who participated in the lecture series in 1852 included the following notable lecturers:

Hon. Horace Greeley, editor of the New York Journal spoke on "The World's Fair."
The Rev. Henry Ward Beecher, "Laws of Precedence"
Rev. John Frederick Schroeder, "Mohammed."
Charles King, "New York History."
Rev. John Pierpont, "Physical Science."
Professor Ormsby MacKnight Mitchell gave three lectures on "Astronomy."

The lecture series continued into 1879 but took a new direction in the ensuing years, changing its course of topics to adhere to the specific interests of the membership working in the labor, landmark and artisan trades. In The Society's annals, the lecture committee predicted:

It is to be hoped that a new impulse has been given to general objects of The Society, and that we shall continue to increase in reputation, and sustain a still higher rank in the cause of benevolence and charity, and the promotion of literature, than we have ever yet done. The advantages which will result from our present efforts will not be confined to our own day, but our successors and future generations will be benefitted by them.

These words of wisdom were fulfilled by the lecture series in years to come.

The Current Lecture Series

The Society's public lecture series, so successfully carried through to the present time (2015), is over 168 years old. As such, it continues the tradition The Society started in 1837. However, the subject matter today diverges into two diverse segments: the Labor, Landmarks and Literature Series and the Artisan Lecture Series.

The Labor, Literature and Landmark lectures present aspects of work and social and cultural history relating to New York City. The goals are to stimulate intellectual exploration, foster aesthetic appreciation and encourage public awareness of the landmark environments, work and culture specific to our city.

The Artisan Lecture Series promotes the work of skilled craftsmen to ensure their unique knowledge is understood and carried forth for future generations. All lectures are presented in the Library's main reading room.

The Labor, Literature and Landmarks Series

In the spring of 2013, the diversification of topics brought additional interesting and provocative Labor, Literature and Landmarks topics to the GSMT platform, and women were in the literature spotlight:

> *Jane Latour spoke about her book "Sisters in the Brotherhood," covering the oral history of women working in non-traditional jobs: an elevator mechanic, and iron worker, a teamster and others.*
>
> *Rachel Bernstein, who for decades focused on working class history in the City, led a discussion on the treasures present in the Labor Arts virtual museum. Bernstein along with the late Debra E. Bernhardt wrote "Ordinary People, Extraordinary Lives: A Pictorial History of Working People in New York."*
>
> *The Cooper-Hewitt Dynasty of New York was the topic presented by the author Polly Guérin, who described the many contributions to the City made by these prominent New York families, including the daughters of Abram Stevens Hewitt (the granddaughters of Peter Cooper) who founded the museum that was renamed in 2014 the Cooper Hewitt, Smithsonian Design Museum.*
>
> *During the 2013 Landmark Lectures New York's official Manhattan Borough Historian, Michael Miscione, a popular lecturer, who enjoys being a historical detective, delivered a talk covering interesting topics about the city.*

Distinguished curators were also invited to produce the labor and landmarks segment of the Tuesday lectures. For example, the spring 2014 landmark lectures were curated by architect Lisa A. Easton, AIA, a partner in the New York City–based architecture and historic preservation firm Easton Architects. Easton brings a font of knowledge to the fore and has worked on the restoration of some of the region's most noteworthy landmarks. Her credentials are numerous, and in addition to receiving multiple awards, she teaches at Pratt Institute Center and lectures extensively at national conferences. During the 2014 landmark lectures, Easton discussed the architectural, social and historical significance of Stanford White's design for New York University's Bronx campus. Three other New York architects highlighted iconic buildings and landmarks of social, historical and cultural significance:

> *Frank J. Prial, Jr., AIA, who is affiliated with Beyer Blinder Belle Architects & Planners LLP, with restoration and revitalization projects including Grand Central Terminal, and the Morgan Library & Museum, spoke about his work on the restoration of the lobby of the Empire State Building in New York City. The work maintained the integrity of the lobby's original design as a testament to the historical significance of the iconic landmark.*
>
> *Kate Lemos McHale, Senior Associate at Beyer Blinder Belle architects, discussed "100 Years Later: New York City's Beaux Arts Masterpieces: Grand Central Terminal, the New York Public Library and The Metropolitan Museum of Art."*
>
> *Cleary Larkin, Associate at Beyer Blinder Belle spoke about the restoration of New York's Beacon Theater.*

Restoring the Mural

Hand craftsmanship illustrated by The General Society's motto "By Hammer and Hand All Arts Do Stand" comes full circle today in the hand craftsmanship work created in the Empire State Building lobby by the award-winning specialty construction, restoration and conservation firm EverGreene Architectural Arts.

Beginning with a historic investigation of the Empire State Building's lobby ceiling, EverGreene designed, created and installed ceiling murals.

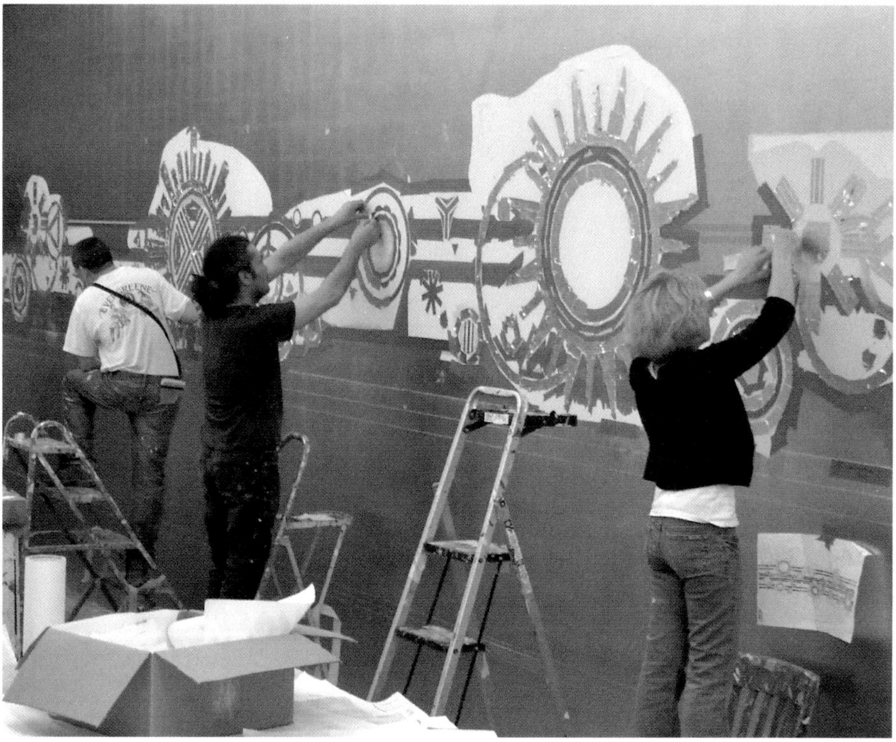

EverGreene conservation technicians and decorative artists restoring the historic Empire State Building mural. *Courtesy of EverGreene Architectural Arts.*

The murals had been obscured and irrevocably damaged by a drop ceiling that was installed almost a half century earlier. To return the lobby interior to its original aesthetic, EverGreene artisans spent two years recreating the fifteen thousand square foot aluminum and gold leaf Art Deco mural. On September 23, 2009, the newly installed mural was unveiled, allowing visitors to experience the Empire State Building as it was meant to be: a glowing tribute to the innovations of the modern age.

Jeff Greene, EverGreene's founder and president, was a guest speaker at the GSMT's Artisan Lecture Series on April 14, 2015. His lecture, "Color, Ornament & Preservation," focused on decorative painting, ornamental plaster and specialty finishes in conservation and restoration projects throughout the United States and abroad.

Behind the Scenes

From time to time, Society members or notable industry leaders coordinated lectures. Such was the case in spring 2014 when five labor components of the lecture series provided a behind-the-scenes look at the creative industries in New York, focusing on the talented artists that deliver the heartbeat of theater, television and cinema. The series was curated in collaboration with Beverly Miller, Local USA 839 president and the United Scenic Artists 829. This series featured:

Inbal Weinberg, "A Day in the life of an Art Director and Production Designer," discussed her career working on numerous feature films and television including the Academy Award nominated films, Frozen River and Blue Valentine, A Place Beyond the Pines and upcoming films including St. Vincent with Bill Murray and Life of Crime with Jennifer Aniston.
Tracy Christensen described "A Day in the Life of a Costume Designer," and her approach to costume research and design for many Broadway and Off-Broadway and regional productions including Twelfth Night; Meet Me in St. Louis, South Pacific and Show Boat, to name a few.
Neil A. Mazzella, "A Day in the Life of a Broadway Technical Director," focused on the multitude of Broadway productions he has provided technical direction for: Bridges of Madison County; Richard III and Twelfth Night; The Glass Menagerie, Breakfast at Tiffany's and Cinderella.
Beowulf Boritt, scenic artist, covered "A Day in the Life of a Set Designer," and spoke about his set design for Broadway productions including the Bronx Bomber; Chaplin; Rock of Ages; Sondheim on Sondheim; The Scottsboro Boys and over fifty Off-Broadway productions.
Thomas Stein, "A Day in the life of a Project Manager," GSMT member (2010), Project Manager of the Shubert Organization Theaters, Stein discussed "Maintaining New York Theaters for Today and Tomorrow." The Shubert Organization has been at the forefront of the American theater since the start of the 20th century and currently owns or operates seventeen Broadway theaters.

The General Society's committee structure, dating back to its founding in 1785, has relied on the members' volunteer work to carry out the many operating needs of the organization. As a member of the GSMT, the aforementioned Thomas Stein is among the members who serve on various committees. Stein has been involved with The Society's school committee and the facilities committee and has filled in as a substitute instructor for electrical technology at the Mechanics Institute.

Labor, Literature and Landmarks Series and the Artisan Lecture Series

The current Labor, Literature and Landmarks Series and Artisan Lecture Series continue the tradition first established in 1837, which stated, "The advantages will result from our present efforts will not be confined to our present day, but our successors and future generations will be benefitted by them." To this end, The General Society honors that dictum.

The lectures bring people with diverse interests from all over the New York area, as well as international visitors, to the GSMT's commodious three-story-high library space on West Forty-fourth Street. As in the past, lectures are usually held on Tuesdays at 6:30 p.m., and just as the Brothers of The Society charged a fee to attend lectures in those early days, the tradition continues today. There is a general admission charge for General Society members and guests and a reduced price for senior citizens and students. The event usually concludes with a reception following each lecture.

From the beginning, supporters and donors have helped to fuel the expense of the lecture series, and today, benevolent and generous support is provided by many sponsors, including the Macquesten Companies, Peter Pennoyer Architects, Ferguson & Shamamian Architects LLP and John B. Murray Architect. The lecture series is also supported, in part, by public funds from the New York City Department of Cultural Affairs, in partnership with the city council.

The Artisan Lecture Series

General Society members have traditionally been instrumental in keeping the lecture series a viable part of the GSMT operations, and in 2010, a new program, the Artisan Lecture Series, introduced noteworthy artisans and their creativity to the inquiring public. This series became a viable part of the Tuesday night lectures and another popular educational and entertaining vehicle initiated by The Society.

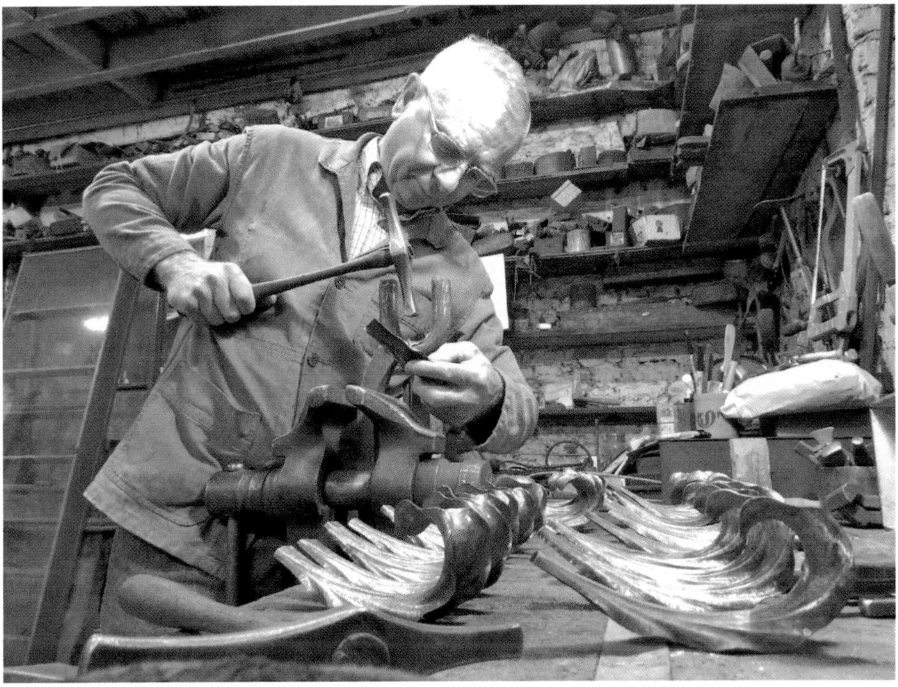

Society member (2009) Jean Wiart at work on one of his creations in his workshop.

Society member (2009) Jean Wiart, a master ornamental metalwork designer and craftsman who oversaw the restoration of the Statue of Liberty's torch and flame in 1993, was instrumental in starting the Tuesday night Artisan lectures in 2010. He selected the best of New York's talent from his vast contacts in the artisan world. The 2011–12 Artisan lectures, for example, featured following artisans:

Pierre Finkelstein, a 1986 Gold Medal graduate of the Van Der Kellen Painting Institute in Brussels, Belgium, wrote the book, "The Art of Faux." He discussed the various techniques used in Trompe L'oeil Painting.

Vanessa Hoheb, who honed her skills in her father's New York sculpture studio, imparted details on the skills and techniques of sculptural enlarging, mold making, casting and restoration.

Bill Gold, owner of the world-renowned lost-wax sculpture foundry, discussed bronze sculpture, statues and monuments.

For the popular Artisan lectures in the 2012–13 season, Jean Wiart recruited celebrated craftsmen to provide detailed descriptions of their art. Master stained-glass artisan Pascale Foucault-Phipps opened the series, followed by master glass engraver Tomas Tisch and master drapery and needle arts artisan Clarissa Barnes DeMuzio.

Additional artisans get their due recognition in the Artisan Lecture Series: Dawn Ladd, lighting artisan, from Aurora Lampworks Incorporated, was featured on May 7, 2013, and spoke about her extensive experience in all things lighting. Her career began when an antique lamp she acquired needed a new shade. Her creation's amazing success launched a commercial career designing custom-made lampshades. Then, as a sculptor, she believed that all objects could potentially be lighting fixtures. Ladd's designs and installations can be seen at the Metropolitan Museum of Art, the Cooper Hewitt, Smithsonian Design Museum, the Eldridge Street Synagogue, the First Presbyterian Church of New York and the Federal Reserve Bank.

When Jean Wiart returned to France to restore and reestablish his father's blacksmith shop, the Artisan Lecture Series was continued under the administration of another Wiart curator. By fall 2013, the lectures were curated by Jean Wiart's daughter, Society member (2012) Camille Wiart. Camille Wiart brings to this curatorship her wide connections among artisans in many diverse fields of craftsmanship. She usually introduces the Artisan lectures at the podium and continues in this volunteer post today.

The artisans in the GSMT's 2014 lecture series paid tribute to the art of craftsmanship by featuring master artisans who lecture about the intricacies of their specialized crafts and creative entrepreneurship:

> *Robert Baird, historical casting master for restoring cast iron buildings, is one of the founders of Historic Arts and Casting, Inc., where he now serves as Director of Business Operations & Marketing and Project Manager. Mr. Baird's discussion shed light on his craft and his company, one of the premier ornamental metal manufacturers in the United States.*
>
> *Sebastian M. Gluck, master organ builder, founder Gluck Pipe Organs, a member of the International Society of Organbuilders and the Professional Circle of the New York Landmarks Conservancy, spoke on the art of organ building and design.*
>
> *Bernard Maisner, Bernard Maisner Calligraphy & Fine Stationery, presented an illustrated lecture which demonstrated the master calligrapher's*

technique. On display were examples of calligraphy and after the lecture the artisan sat down and with his quill pen created some guest's names in calligraphy flourishes of loops and swirls.

The ongoing Artisan Lecture Series, along with the Labor, Literature and Landmarks Lectures, stands as a testament of tribute to The General Society's first course of lectures established in 1837 that drew in the citizens of Old New York who were eager for self-education. The current lectures continue to provide the same purpose, to be a platform for enlightenment, education and entertainment at the GSMT headquarters.

THE GSMT'S SMALL PRESS CENTER/NEW YORK CENTER FOR INDEPENDENT PUBLISHING

Throughout The General Society's history, the organization has been a forerunner of innovative programming and has its roots in education, lectures and literature, commencing in 1837 when speakers were engaged to provide diverse topics of interest to the Mechanics School students, their families and the inquiring public.

The General Society building, located in the culturally rich neighborhood of Literary Row on West Forty-fourth Street, was the ideal setting for yet another venture: The New York Center for Independent Publishing (NYCIP), which was founded as the Small Press Center:

Strolling down forty-fourth street one day, Society member (1996) Whitney North Seymour, Jr., a lawyer and former New York State Senator, was impressed with the GSMT's building and its showcase windows at street level. As a small press publisher he recognized the potential of The General Society Library as a possible home to assist the nation's small presses. The General Society responded positively to this idea and so the Small Press Center (SPC) became effectively operative in 1984 to encourage writers and poets and to assist independent small publishers who make literature come alive as printed books.

The SPC grew from its early beginnings as a shelf of books and an occasional program to an educational organization that encompassed a several thousand–strong library on publishing and a display of hundreds of small press titles. For many years, the Center ran fifty-plus publishing, writing and literary programs and was also the main sponsor for Small Press Book Fair, the annual New York Round Table Writers' Conference and one of the cosponsors of the national initiative to help independent presses, Small Press Month. At one time, approximately one thousand recent books published by small presses were on display in The General Society library.

SMALL PRESS AUTHORS

Among the authors who have turned to small press publication in the past are the Bronte sisters, Emily Dickinson and Marcel Proust, all of whose works were turned down by mainstream publishers. Still, these authors were honored and remembered. For instance, at the ninth annual Small Press Book Fair, the focus was on the accomplishments of Virginia and Leonard Woolf and the publishing activities of Hogarth Press and how this is happening in the independent publishing world today.

Poet Marianne Moore was featured on the cover of the 1989 Book Fair program as that year's example of an author who became important for her literary contributions after first appearing in small press publications.

Walt Whitman was honored as the Small Press Center's Author of the Year, 1990, with an exhibit of the first edition of *Leaves of Grass* and other memorabilia. The archives of the GSMT also mention Ralph Waldo Emerson, who was honored and remembered.

Nearly three thousand visitors attended that year's Book Fair and viewed books published by two hundred small presses. Well-known contemporary authors who were published by small presses participated in the center's programs, such as Small Press Month, including Alice Walker, Kurt Vonnegut and Sherman Alexie.

Awards and recognition for Small Presses figured importantly in other ceremonies. Prior to the 15th Annual Small Press Book Fair in 2003, Ragtime *author E.L. Doctorow presented George Plimpton, editor/publisher of the* Paris Review *with a small brown bust of Benjamin Franklin—the Poor Richard Award—for Outstanding Achievement in Independent Publishing. At that time Plimpton said, "The* Paris Review *is fifty years old and still alive, though occasionally it has teetered on the verge of extinction. Literary magazines are like butterflies. They usually don't last very long."*

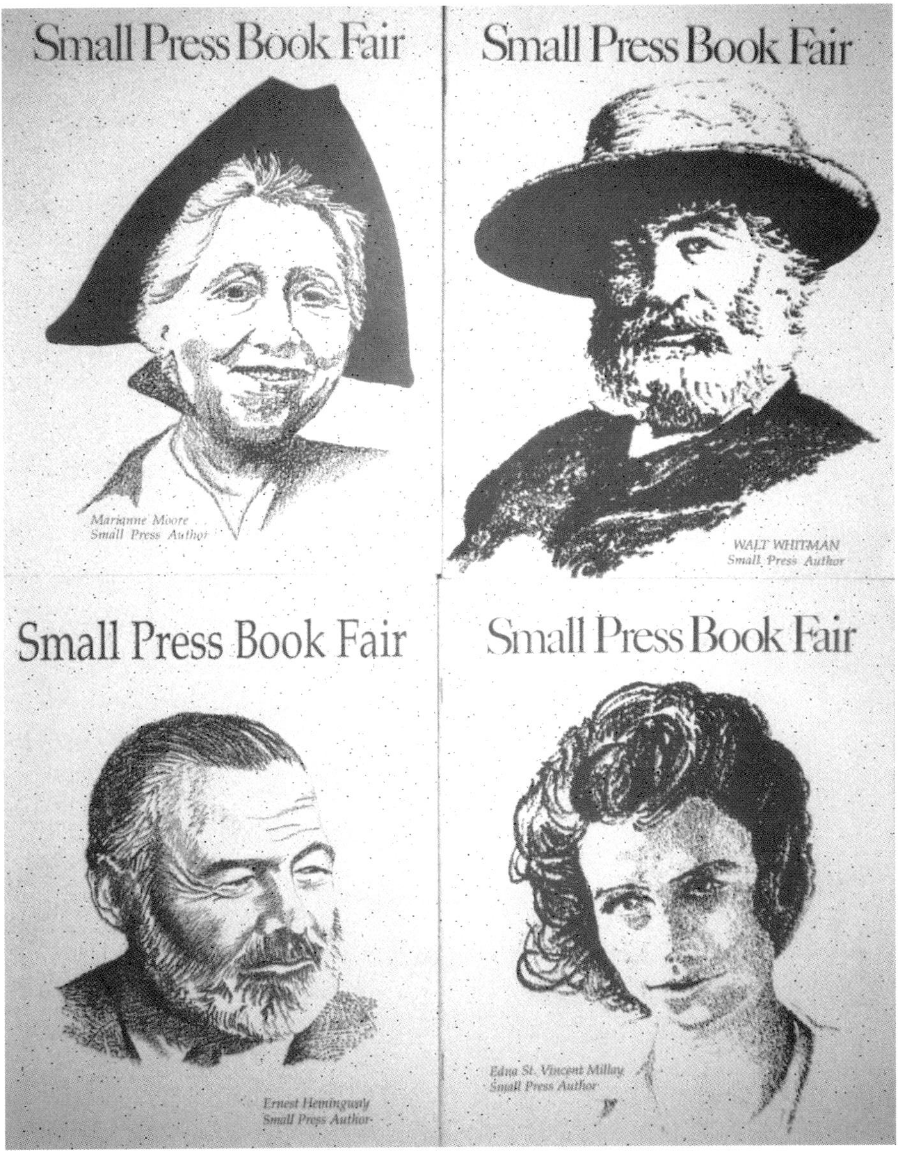

Montage of cover art for Small Press Book Fairs, 1989–97. *Left to right*: Marianne Moore, Walt Whitman, Ernest Hemingway and Edna St. Vincent Millay. Artist: Catryna Ten Eyck Seymour.

The Small Press Center's objective focused on the diversity of independent publishing—from books on literature to self-help books, from cookbooks to history books—and provided a wide range of books and authors that were otherwise not generally available. SPC's impact was not limited to the United States and served as a prototype for a companion organization in Great Britain and a like-minded group in Canada. Case in point: an international aura pervaded the 1989 Small Press Book Fair, which had exhibits from the British Small Press Group in London; Small Press Network in Vancouver, B.C., Canada; and Literary Press Group of Canada from Toronto; as well as books from China, Ireland, Africa, Russia and other countries.

It is interesting to note that The General Society's historical role as the seedbed of independent publishing dates back even earlier, to 1831, when James Harper—one of the original four Harper brothers, publishing giants—joined as an active member of the GSMT, which has housed the lectures and literature center since its inception.

Name Change to NYCIP

The Small Press Center changed its name to the New York Center for Independent Publishing in 2007 to designate the full scope of its support of the independent publishing sector and continued the tradition of book fairs and writers' conferences and cosponsored Small Press Month.

For the ensuing years, educational programs and books remained the NYCIP's main focus. Among its programs, the NYCIP delivered a series of workshops and lectures led by outstanding publishing personnel to help writers, aspiring publishers, marketing staffs and anyone intrigued with the challenge of entrepreneurial publishing.

Literature at The General Society

The recent literature component of The General Society's Labor, Literature and Landmarks Series focuses on publications and periodicals intrinsic to the culture of New York City. Featured publications included the *New York Times*, the *New Yorker*, *Harper's* and the *Nation*. Writers and editors who have spoken as part of the series include David Remnick, Adam Gopnik, Hendrik Hertzberg and Bob Manoff of the *New Yorker* and Sam Sifton, Jill Abramson, Ben Brantley and Arthur Gelb of the *New York Times*.

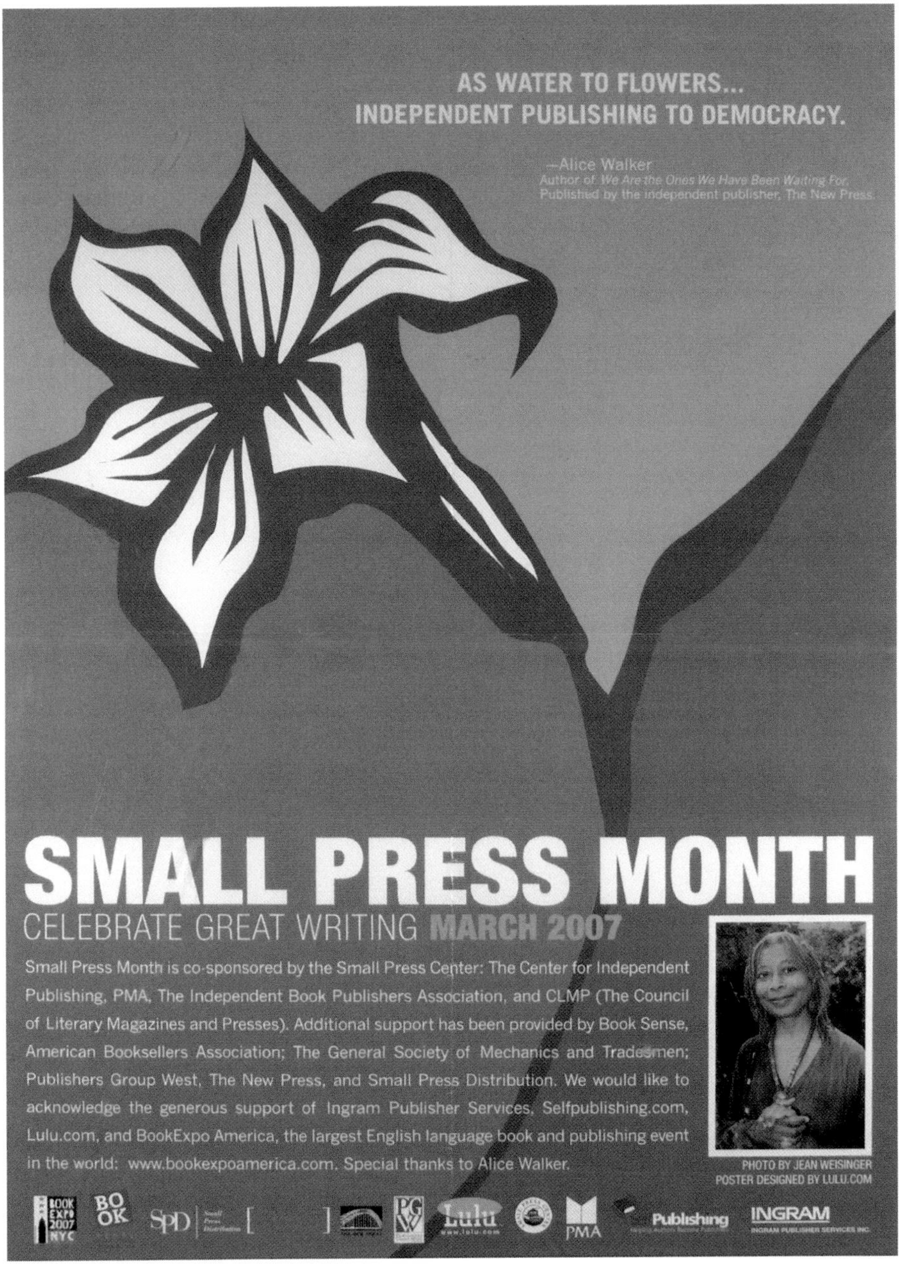

Alice Walker, noted Pulitzer author of *The Color Purple*, honored during Small Press Month, 2007. *Poster design by Lulu.com. Photograph by Jean Weisinger.*

The list of prestigious writers who have spoken at the GSMT under the umbrella of the Small Press Center/the New York Center for Independent Publishing included Tom Wolfe, Alice Walker, Richard Ford, Michael Cunningham, Walter Mosley, Mary Higgins Clark, George Plimpton, Michael Connelly, Meg Wolitzer, Lincoln Child, Nuala O'Faolain, Sigrid Nunez, Alice Hoffman, John Berendt, Joshua Ferris, Richard Price, Colson Whitehead, Jonathan Ames and E.L. Doctorow.

On the occasion of the GSMT's 200th anniversary, Brendan Gill, theater critic for the *New Yorker* and chairman emeritus of the New York Landmarks Conservancy, offered a lecture about the illustrious Literary Row neighborhood, "A Palimpsest of Wonder: Lively Days and Nights on West Forty-Fourth Street." It was a fitting tribute to the GSMT, which, in its 230th year, remains the epicenter of lectures on West Forty-fourth Street.

Today the GSMT moves ever forward in its quest to provide New Yorkers with innovative programs and, to this end, continues to bring publishers to the lecture platform. In 2014, the GSMT invited Pegasus Books, one of the fastest-growing independent publishing companies in the country, to present an insider's view of independent publishing. Like Pegasus Books, the GSMT

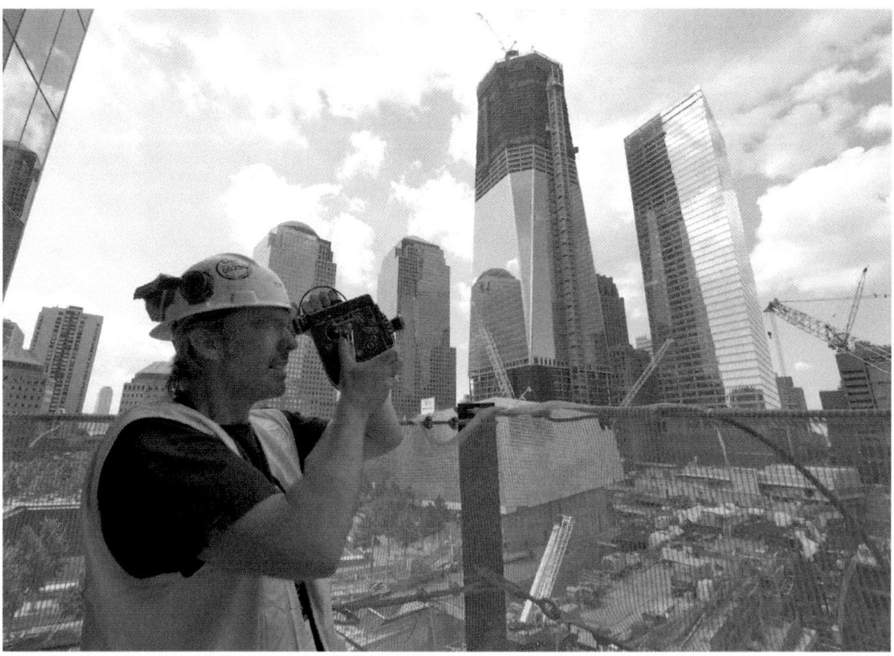

Marcus Robinson recording his award-winning film *Rebuilding the World Trade Center*, presented at the GSMT on Tuesday, October 7, 2014. *Photograph by Allan Tannenbaum.*

shares in the conviction that good literature is essential to the health of our cultural life.

Taken as a whole, the GSMT shines a bright light on the creative resources of the city by revealing how skilled authors and lecturers contribute to the ongoing vitality that makes New York City the literary capital of the country.

The Society is always cognizant of significant historical events in New York City and continues to speak to the modern seeker of knowledge through innovative venues. For example, the GSMT was honored to present, on October 7, 2014, a showing of Irish artist/photographer Marcus Robinson's award-winning documentary film *Rebuilding the World Trade Center*, which was presented in the Library. That a record audience of New Yorkers came to view this unique film bears witness to its historical significance. Robinson said, "Out of the most devastating trauma, something amazing can be rebuilt."

Rebuilding the World Trade Center tells the story of architecture and engineering but focuses on the hero laborers, from the site managers to the legendary ironworkers who assembled the steel frames on girders high in the air. For the past ten years, Robinson has been at the World Trade Center in lower Manhattan, filming, photographing and painting the transformation of the historic World Trade Center site. With thirteen time-lapse cameras running, Robinson condensed years into a few seconds, capturing vast buildings rising over the city.

His film is an epic story about the transformation of the site and the construction of the towering buildings. Most notably, it celebrates the diversity of the workers, whose commentary in the film brings to life the hero workers who continue to build our magnificent city.

The General Society programs and lectures reflect the cultural breadth of New York City and continue to be as relevant, informative and entertaining in 2015 as they were in 1837.

11
The GSMT Awards

The basic goal of labor will not change. It is, and it has always been and I am sure always will be, to better the standards of life for all who work for wages and to seek decency and justice for all Americans."
—George Meany, president of the AFL-CIO, 1955–1979.

The General Society of Mechanics and Tradesmen stands firmly anchored as the preeminent organization that recognizes those members in the construction industry who continue to build a better future for all New Yorkers. To this end, the Craftsmanship Award is presented each year to individuals in the building and construction trades who have demonstrated leadership in their fields; promote high professional standards; and display pride, skill and quality in their work.

The award ceremony is an annual event, which was initiated in 2004 and is given to honor individuals who symbolize the best in craftsmanship in New York City. It also honors those corporations who help to secure the future of the Mechanics Institute through their generosity so the Institute can continue its long tradition of serving the building and construction industry.

In thanking donors at the 2014 Craftsmanship Awards, Society member (1992) Gerard Drohan Jr.—past president (2007–08), vice president (2013–14) and president (2015)—said:

> *Each year, as it has done since 1858, the Mechanics Institute opens its doors to students possessing a desire to advance themselves in their chosen field. Since*

its beginning the Institute has trained more workers for the industry than any other institution in the nation. Our graduates are active in many areas of New York City's Building and Construction Industry and many have gone on to attain prominent positions within their field. The Institute offers its comprehensive instruction to qualified persons on a tuition-free basis and is financed through contributions to The General Society.

THE CRAFTSMANSHIP AWARDS

The GSMT is justly proud of its close association with industry leaders and is always on the cusp of innovation. The recipients of the Craftsmanship Awards receive an exquisite artifact of appreciation, an award created by one of The Society's own members.

Before Society member (2009) Jean Wiart departed for France, in his atelier, he designed the laurel leaf award for the 2011 Craftsmanship Award honorees. It is sculpted from wrought iron with gold leaf covering the leaves and symbolizes the fact that the meritorious recognition given to craftsmanship

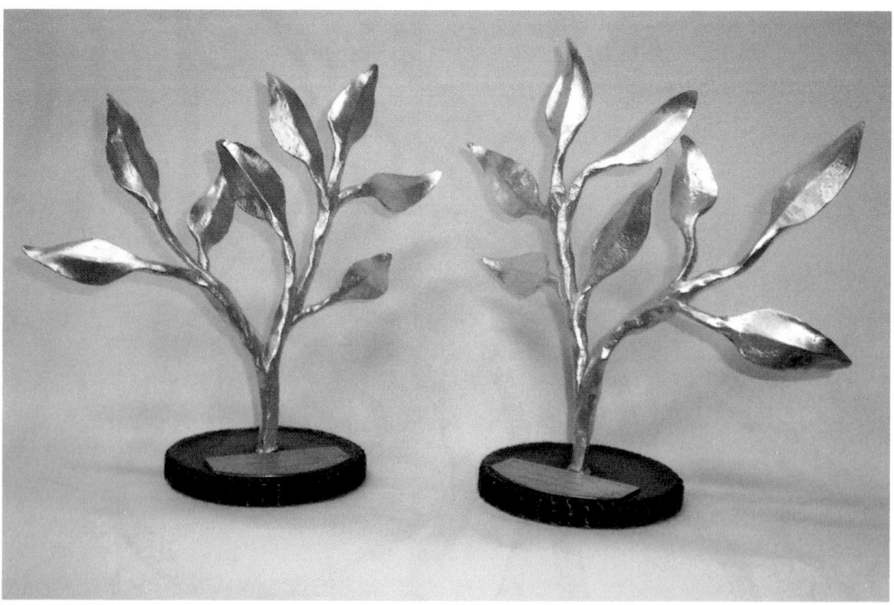

Gold Leaf Award designed by GSMT member Jean Wiart, sculpted in wrought iron with gold leaf covering the leaves.

honorees is a symbol of glory, just as the ancient laurel leaf was for both Greek Olympic Game winners and Rome's victorious generals.

Society member (2003) Lenore Janis, president of Professional Women in Construction, received the 2011 Craftsmanship Award, which recognized her long career in the construction industry and contributions in encouraging women to enter the field.

At the annual Craftsmanship Awards benefiting the Mechanics Institute, Charles D. Avolio, LEED AP, vice president and operations manager, Turner Construction Company, received the 2012 Craftsmanship Award. In addition to Avolio's managerial positions at Turner Construction, he also serves as principal in charge for several of Turner's ongoing projects, including 2 World Trade Center, the John Jay College Expansion Project and St. Mary's Hospital for Children.

Society member (2012) Jay Badame—president and chief operating officer of New York, New Jersey and Pennsylvania for Tishman Construction, an AECOM company—presented the 2012 award to Avolio.

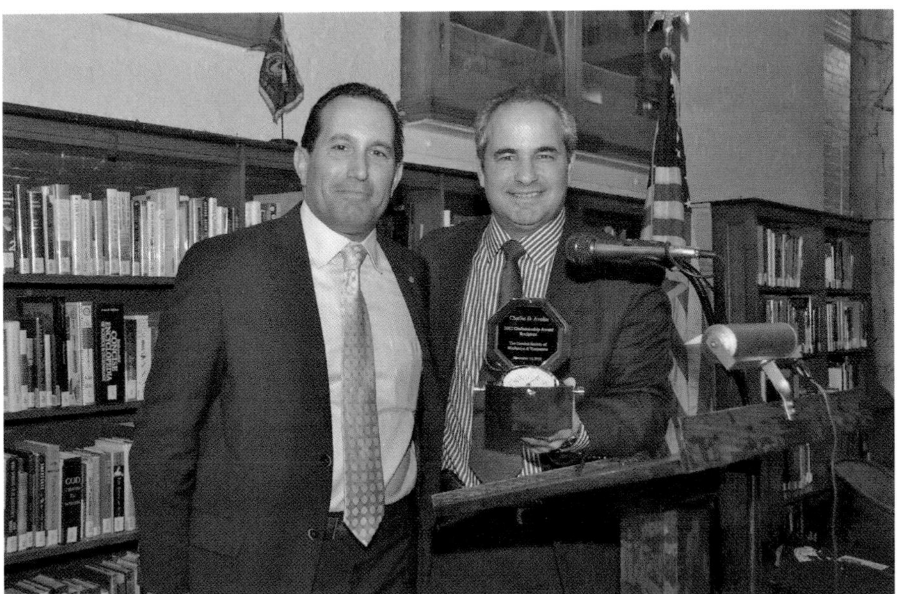

Jay Badame, president and chief operating officer of Tishman Construction Corporation, presents Charles D. Avolio, of Turner Construction Company, with the 2012 Craftsmanship Award. *Photograph by D.A. Luhmann.*

It is interesting to note that Jay Badame was a recipient of the 2008 Craftsmanship Award. Under his leadership, Tishman expanded its capabilities in the high-rise residential market and continues to be a leading builder in the New York/New Jersey region. During his career, Badame has led his company's efforts on many of New York City's most iconic structures, including One World Trade Center, One Bryant Park, Hudson Yards, Carnegie Hall, the Jacob K. Javits Convention Center and the New York Stock Exchange.

AVOLIO'S SPEECH

At the 2012 award ceremonies, Avolio eloquently spoke of the Mechanics Institute's mission. Following is an excerpt from his speech:

> *Every person involved with a construction site has a contribution to make. Many tradesmen have the skills and the talents to put the building together with their hands, but others who are not very good with their hands, provide other skills to manage a team of people to build a project as a designer envisioned it, and to the expectations of our clients. The education that folks get at the Institute has a direct contribution to the industry, not only in the skills that are learned, but in other intangible ways like instilling accountability, responsibility, teamwork and accomplishment—all attributes that are essential in our business. Training and education of those who will follow after us is of great importance if we are to grow and improve this industry. Part of my job is to manage personnel and their careers and staff the projects. At Turner there is always the charge: "Train your replacement," and "that is what goes on here at the Institute."*

Society member (1978) Viggo Bech Rambusch was a 2007 Craftsmanship Award recipient and is the honorary chairman and senior product manager at Rambusch Design Studio/Workshop. The rich legacy of the Rambusch Company spans more than a century of architectural history and is prominent in projects around the world.

The World Trade Center

It is worthy to include that, in the same year, Larry A. Silverstein, president and chief executive officer of Silverstein Properties and the developer of the new World Trade Center, was also a recipient of the 2008 Craftsmanship Award.

A historic event is forever linked to Silverstein Properties. In July 2001, Silverstein had completed the largest real estate transaction in New York history when he signed a ninety-nine-year lease on the 10.6-million-square-foot World Trade Center only to see it destroyed in terrorist attacks six weeks later on September 11, 2001.

Rising from the ashes, Silverstein's company went on to rebuild the $7 billion office component of the World Trade Center site, and in May 2006, Silverstein opened 7 World Trade Center, a fifty-two-story office tower at 250 Greenwich Street. In September of the same year, he unveiled designs for three other office towers on the site, each designed by Pritzker prize–winning architects. In November 2013, 4 World Trade Center opened and 1 World Trade Center, the tallest building in the Western hemisphere, opened in October 2014.

It is fitting that the World Trade Center properties developer, Larry A. Silverstein, and no other magnate of his stature, could describe the colossal rebuilding of the World Trade Center. Silverstein's remarks are recorded in Marcus Robinson's *Rebuilding the World Trade Center* film.

Praise for Mechanics Institute

The General Society's 2013 Craftsmanship Award was presented to Michael Della Rocca, at that time the North America, chief executive of the AECOM Technology Corporation to celebrate his career as an industry leader, which spanned more than thirty years.

In his acceptance remarks, Della Rocca acknowledged the nearly two and half centuries of contributions the GSMT had made to the city of New York in carrying out its mission:

> *To serve and improve the quality of life of the people of the City of New York through educational, philanthropic and cultural programs.*

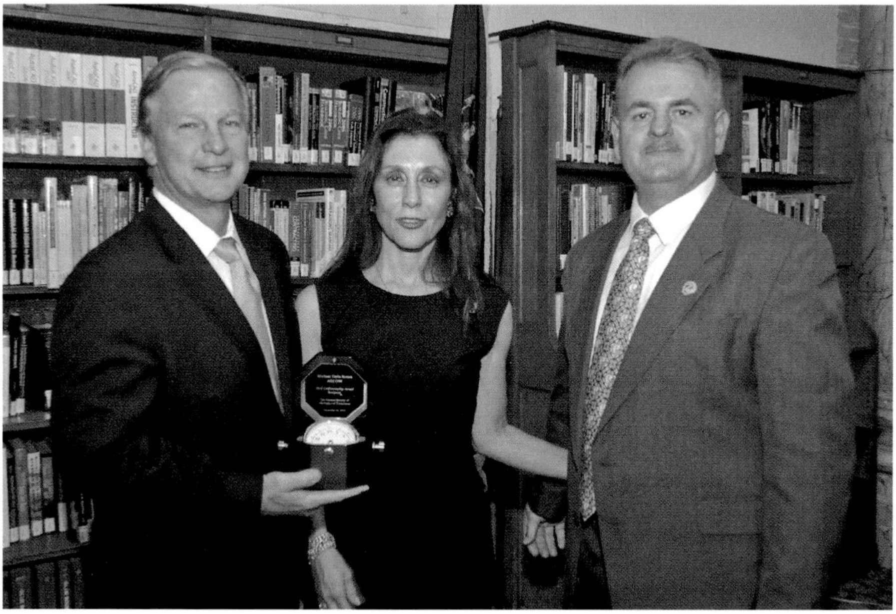

Michael Della Rocca accepting the 2013 Craftsmanship Award from executive director Victoria A. Dengel and GSMT president Gottfried Weissgerber. *Photograph by D.A. Luhmann.*

He made a point to acknowledge that AECOM and the entire construction industry in the city had benefited from the skills many of their workers had learned at the Mechanics Institute.

THE 2014 AWARDS

The Annual Craftsmanship and Civic Leadership Awards celebration reached its zenith on November 20, 2014, when a kilt-costumed bagpiper greeted the attendees as they entered the Library. One may assume that this musical interlude was perhaps in reverence to the Scotsman Andrew Carnegie, whose generosity made the Library's 1903 extension possible.

The Craftsmanship Award Ceremonies, benefiting as it does the Mechanics Institute, honored Staten Island native Frank Gibbons, chief operating officer of Penguin, a wholly owned subsidiary of the EMCOR Group. Gibbons, a Society member and an alumnus of the Mechanics Institute, is instrumental in driving the philanthropic goals of giving back to the community and has led Penguin's efforts in support of EMCOR's philanthropic initiatives, including hosting Pink Hard Hat events for Memorial Sloan Kettering and Gildas Club of New York City.

At the same celebration, New York's Fourth District councilman Daniel R. Garodnick received The General Society's Civic Leadership Award for 2014. Garodnick's newly designed District Four now includes The General Society's building on West Forty-fourth Street. Most significantly, he has been at the forefront of many community issues and active in all significant district events.

Civic Leadership Awards

The General Society's Civic Leadership Award, initiated in 2012, was given in that year to Donald Rubin, co-founder and co-chair of the Rubin Museum of Art, which opened in October 2004. He and his wife, Shelly Rubin, have been collecting Himalayan art for more than thirty years; their treasured collection became the basis of the museum. The Rubins have also contributed widely to health, human rights and culture, with special attention paid to philanthropy that is strategically effective. In accepting the award, Donald Rubin acknowledged the contribution the Mechanics Institute has made in improving the education of so many students in the construction industry who help to build and redefine the constantly changing skyline of New York City.

In the same year, the posthumous Civic Leadership Award honored Jay Rubin, president emeritus of the New York Hotel and Motel Trades Council, AFL-CIO, for his years of service to the Labor Movement. Peter Ward, president of the New York Hotel and Motel Trades Council, accepted the posthumous honor. Jay Rubin was the chief figure in a 1937 drive to organize twenty-five thousand hotel workers, and in 1939, he became president of the New York Hotel Trades Council, which negotiated and enforced the city-wide contract covering hotel workers who had worked ten- and twelve-hour days, seven days a week at very low wages and no benefits. Ward stressed the groundbreaking importance of Jay Rubin's forty years of union leadership and recalled Jay Rubin's words, "We brought dignity to one of the most depressed and exploited categories of workers."

In 2013, the Civic Leadership Award went to the thirty-second New York City fire commissioner, Salvatore J. Cassano. In accepting the award, Commissioner Cassano thanked the Mechanics Institute:

For the work it does to train, so many members of the construction workforce in the City and for the safe, quality work the Mechanics Institute graduates perform, and also noted that the valuable skills of mechanics and tradesmen, performed vital tasks following Hurricane Sandy's destructive surge.

It is interesting to note that Jack Lerch, honorary chief of the fire department of New York at the George F. Mand Library, said, "Cassano was one of the greatest fire commissioners that New York City ever had. As chief of Operations and later Chief of the FDNY Department, Commissioner Cassano played a crucial role in rebuilding the Department in the wake of September 11 and the loss of 343 members of the Department. More than 6,000 new firefighters were hired and 3,000 others promoted to various officer ranks as the Department went through a difficult but remarkable resurgence."

The General Society's awards also extend to individuals in the community who have made significant contributions to New York City. In 2011, the Civic Leadership Award was given to the supermarket magnate and Greek immigrant John Catsimatidis, owner and chief executive officer of Red Apple Group, which includes Gristedes and Red Apple supermarkets as well as oil and real estate holdings. The award recognized Catsimatidis for his many contributions to New York City's economic development and to a wide variety of health and educational organizations. He was just six years old when he became a United States citizen and has lived the American dream as an industry leader.

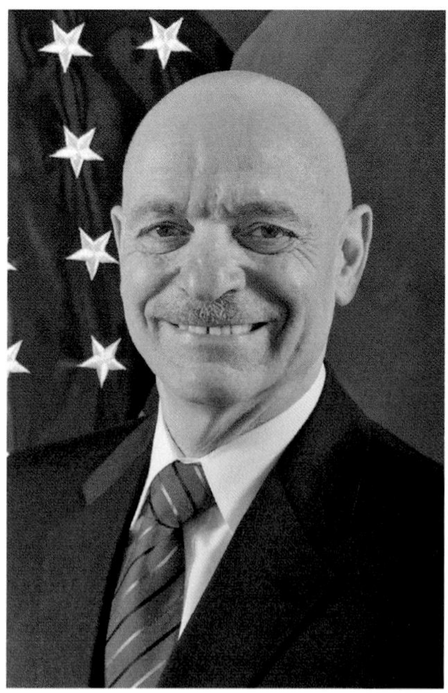

New York fire commissioner Salvatore J. Cassano received the 2013 Civic Leadership Award.

Proceeds from the event, held at the Union League Club, benefited the GSMT's Mechanics Institute. Major contributors included Margo and John Catsimatidis as well as Rella Fogliano-Macquesten Construction Management, LLC; John Gallin and Son, Incorporated; Daniel Rose, Rose Associates; and Jack and Susan Rudin.

Excellence in Building Technology

The drumbeat of awards and recognition given to industry innovators goes on. Technology has become a vital tool for the building industry, and in 2013, The General Society acknowledged its importance by creating a new award, Excellence in Building Technology. Construction of major projects today requires the latest in Building Information Modeling (BIM)

State of New York
Executive Chamber
Albany 12224

Andrew M. Cuomo
Governor

November 14, 2013

The General Society of Mechanics & Tradesmen
20 West 44th Street
New York, NY 10036

Dear Friends:

It is a pleasure to send greetings to everyone gathered for The General Society Annual Craftsmanship and Civic Leadership Award.

We recognize the valuable role of organizations that function as essential resources for the working men and women who have defined the strength and character of our great nation through their labor. Since 1785, The General Society has furthered the building and construction industry in New York City, creating a trained workforce, and a path to advancement in various trades, through tuition-free education and development.

Tonight, you come together to pay tribute to those who promote exceptional professional standards and I join in congratulating your distinguished honorees. On behalf of all New Yorkers, I commend the leadership and membership of The General Society for its proven dedication to providing service to the people of the City of New York.

With warmest regards, and best wishes for an enjoyable evening.

Sincerely,

ANDREW M. CUOMO

We Work For the People
Performance ★ Integrity ★ Pride

printed on recycled paper

Governor Andrew M. Cuomo's congratulatory letter on the occasion of the Annual Craftsmanship Award and Benefit on November 14, 2013. *Courtesy of the* Commemorative Journal.

technology. For example, Autodesk, Incorporated, a leader in developing this technology, provides a comprehensive portfolio of building solutions for both CAD and BIM workflows that allow architects, designers and contractors to create compelling visualizations of a building's progress that becomes a permanent record. Martin Beahm, Autodesk's New York City representative, congratulated the Mechanics Institute for now making this technology available to the Institute's students and accepted the award for his company.

Honoring the GSMT Ancestors

Although the Mechanics Institute is proud of all its graduates, The Society's historical records maintain references to some of their most celebrated graduates.

For instance, there is Charles DeBenedittis, senior managing director and senior advisor of design and construction, Tishman Speyer Properties, Incorporated, which is part of the vast construction industry that has built New York into the great metropolis it is today. He is justly proud of his heritage and honors the memory of his father, Anthony DeBenedittis, who attended classes at the Mechanics Institute from 1924 to 1927 (graduated 1927) and was always proud to display his diploma. The elder DeBenedittis's story is typical of the time. As an immigrant, he arrived from Italy in 1920 at the age of sixteen. With his training at the Mechanics Institute, he went on to become a successful general contractor and developer.

Charles DeBeneditttis also has historical pride in the fact that he was involved in the initial Port Authority negotiations and trade awards at Tishman Realty for the original World Trade Center when it was constructed a generation ago. Carrying on the DeBenedittis name in the construction industry and following in the footsteps of the family's tradition, DeBenedittis's grandson, Eric, is now working for Turner Construction Corporation at the World Trade Center.

Then, too, there is Vanessa Hoheb, the maternal granddaughter of Anthony Moschella, who recalled that her grandfather attended the Mechanics Institute from 1927 to 1930 and graduated in 1930 with a degree in architectural drafting. When he came to the United States, a well-connected family, the Tricomis, took him in, and Mary Tricomi enrolled

Anthony in the Mechanics Institute. Later, he was hired by the family's construction firm, Tricomi Construction Company, and became a master bricklayer, construction foreman and, afterward, began his own company where he taught Hoheb's father, Bruce Hoheb, the masonry trade.

This story has historical significance as the artisan talent was born again with Vanessa Hoheb, who is a master sculpture enlarger and modeling artisan. She honed her skills in her father's New York sculpture studio where she learned the techniques of sculptural enlarging, mold making, casting and restoration. Hoheb is in demand for her craft. In 1968, for example, she headed the five-member team responsible for restoring the skin of the Statue of Liberty. Hoheb has also restored many memorials and monuments and taught mold making and casting at the Pratt Institute.

Another scion of the construction industry and longtime supporter of The General Society's Mechanics Institute is Jonathan D. Resnick, president Jack Resnick & Sons, owners and builders since 1928. Jonathan remembered his grandfather Jack Resnick who was enrolled in the 1930s in the plumbing course at the Mechanics Institute. The GSMT archives provide copies of Jack Resnick's class attendance records and an original copy of his graduation program.

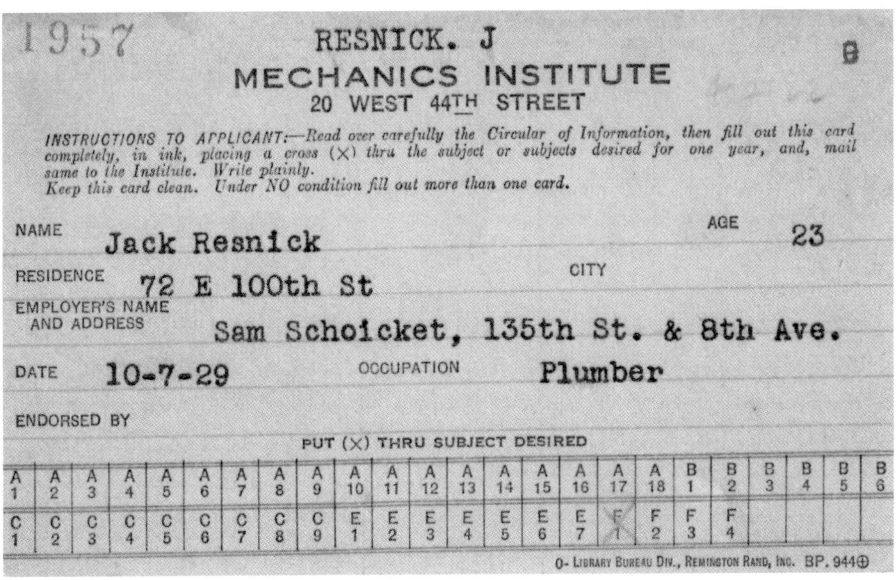

1957 RESNICK. J B

MECHANICS INSTITUTE

20 WEST 44TH STREET

INSTRUCTIONS TO APPLICANT:—Read over carefully the Circular of Information, then fill out this card completely, in ink, placing a cross (X) thru the subject or subjects desired for one year, and, mail same to the Institute. Write plainly. Keep this card clean. Under NO condition fill out more than one card.

NAME Jack Resnick AGE 23

RESIDENCE 72 E 100th St CITY

EMPLOYER'S NAME AND ADDRESS Sam Schoicket, 135th St. & 8th Ave.

DATE 10-7-29 OCCUPATION Plumber

ENDORSED BY

PUT (X) THRU SUBJECT DESIRED

A 1	A 2	A 3	A 4	A 5	A 6	A 7	A 8	A 9	A 10	A 11	A 12	A 13	A 14	A 15	A 16	A 17	A 18	B 1	B 2	B 3	B 4	B 5	B 6
C 1	C 2	C 3	C 4	C 5	C 6	C 7	C 8	C 9	E 1	E 2	E 3	E 4	E 5	E 6	E 7	F 1	F 2	F 3	F 4				

O- LIBRARY BUREAU DIV., REMINGTON RAND, INC. BP. 944⊕

Jonathan Resnick of Jack Resnick and Sons remembers his grandfather Jack Resnick, Mechanics Institute diploma recipient, plumbing, 1930. Pictured here is Jack Resnick's MI Registration card.

The ancestral connections to the GSMT are rather interesting because The Society seemingly has had a profound influence on family members who continue to be affiliated to the organization.

For instance, Society member (1992) Gerard Drohan Jr.—president (2007–08), vice president (2013–14) and president again in 2015—has a special bond with the GSMT through his father, Society member (1972) Gerard Drohan Sr., who served with distinction as president of the GSMT (1988), and from 1972 to 2006, he was an active member of The Society and served on numerous committees. Drohan Sr. graduated from the Mechanics Institute in 1961 and was an instructor at the Institute for decades and included among his affiliations membership in the Mechanics Institute Alumni Association.

Taking this connection a step further, the Drohan family's relationship with The Society is longstanding. Gerard Drohan Sr., for example, is the grandfather of GSMT employee Margaret Stanton, who joined The Society in 2014 to fulfill the position of registrar of the Mechanics Institute. She also serves as The Society's program assistant. Adding to this interesting connection is but one more clue: Stanton's uncle is Gerard Drohan Jr.

A third generation GSMT member (1942), Alexander S. Walker now lives in Nevada, but he still has strong ties to The Society. His father, Alexander C. Walker, was a member in 1912 and served as president in 1927. In a gesture of benevolence, Alexander S. Walker's grandfather, Society member (1894) and president (1912) Alexander Walker, had bequeathed the sum of $25,000 to The Society, the principal of which is known as the "Alexander Walker Fund" and the income to be used for the general purposes of The Society.

Most significant among family ties to the GSMT is Victoria A. Dengel, the current executive director of the GSMT, whose father, Gerard A. Dengel, was president of The Society in 1982. Her role as executive director is discussed in the next chapter.

There is something—as the French say, "*Je ne sais quoi*!"—about The General Society of Mechanics and Tradesmen that seems to draw individuals who desire to be affiliated in some way with the GSMT through family ties or professional interest. In doing so, these men and women are members of the new generation who will continue to uphold the organization to its highest merit in the city of New York.

12

The General Society Then and Now

Withall, I believe it can be said that the interest of the artisan, the mechanic and the craftsman and the commitment to excellence that he represents will always be part of any future.
—Gerard A. Dengel

The General Society of Mechanics and Tradesmen holds an honorary place in New York City's sense of history and civic pride. "And still we are uncertain, and perhaps more uncertain of the exact nature of the future than were our predecessors in 1785." These remarks by Society member (1972) Gerard A. Dengel, president (1982), resonate even today. In a welcoming address at an Independence Day Celebration in 1982, President Dengel said:

> *With the "Declaration" as reference, the Founding Fathers did, then, create a structure of government in which the evolutionary process was made organic. Today the single, most important characteristic of the American Democracy—economic, social and political—is that it is evolutionary—rather than revolutionary. That it should remain so is part of the task of each succeeding generation of Americans: part of our, patient, task.*

The building of the nation can be attributed to the laborers of the construction industry to which The General Society of Mechanics and Tradesmen is so closely aligned, and the organization's motto "By Hammer and Hand All Arts Do Stand" is a firm tribute to the building trades.

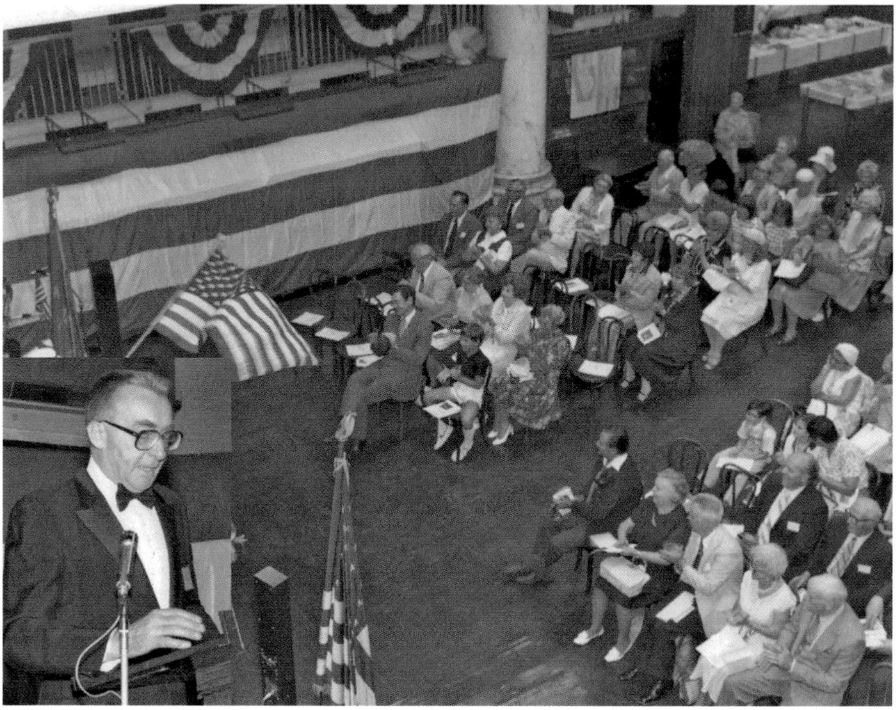

Gerard A. Dengel, GSMT president (1982), at the podium. Welcoming address at the 1982 Independence Day Celebration, July 7, 1982.

Throughout the years, the GSMT has maintained a steady course honoring the history of The Society. Through its educational facility and its outreach programs—including the Library, the lecture series and cultural activities—the organization has also contributed exponentially to New York City's development as a major cultural center of the world.

On its uninterrupted path into the future, the GSMT (1785–2015) celebrates 230 years of continuous service that has genuine resonance for the people of New York City. As such, The Society, founded in the early days of the Republic, has made history-changing advancements in education. Since 1820, the GSMT has enriched the lives of students through its school, which became the Mechanics Institute in 1858. It is the only institute of its kind in the United States providing tuition-free technical education to men and women in the building trades. Today, the Institute is ahead of trends with technology and offers state-of-the-art disciplines and advanced studies in high-tech developments, including CAD and BIM classes.

The multi-level General Society Library, a unique sanctuary in the city, is the destination where scholars, students and the general public conduct study and research. Founded in 1820 as the Apprentices' Library, an educational resource for apprentices and factory workers, it was the second-oldest library in New York City. The General Society Library, as it is called today, has reinvented itself as a cultural and educational center. As such, it has become an active facility in the culturally rich neighborhood located at 20 West Forty-fourth Street. In addition, The Society's street-level showcase windows are a distinct attraction that catches the attention of the passing public and creates much interest in the Mechanics Institute and the Library.

Patriotic Pride

The GSMT on West Forty-fourth Street is located right in the heart of Clubhouse Row, and it is one of the most patriotic venues in the city. The street is distinguished by its impressive display of American and clubhouse flags unfurled together outside almost every building between Fifth and Sixth Avenues. Heading west from Fifth Avenue, the first flag to attract attention is the GSMT's royal blue flag, imprinted with The Society's motto "By Hammer and Hand All Arts Do Stand." It is interesting to note that, in 1988, The Society's building façade was designated a New York City Landmark, and in 2008, the building was placed on the National Register of Historic Places.

The flags on Clubhouse Row also adorn the Harvard Club, New York Yacht Club, Association of the Bar of the City of New York, Penn Club and the Hotel Algonquin, where the fabled Round Table of literary wits once reigned supreme in the intellectual and social circles of the city.

Groundbreaking Lectures

The Society's original lecture series was also an active part of the literary scene of an earlier era. However, unlike the demise of the Algonquin Round Table, The General Society continues to bring the popular Labor, Literature and Landmark Lecture Series and the Artisan Lecture Series to the forefront of public interest, and thereby contributes to the cultural enrichment of

New York. The ongoing public lecture series are a brilliant facet chiseled into history of The General Society, harkening back to 1837 when lectures were first established, just fifty-two years after The Society's founding.

Today, The Society's lecture series consist of compelling interdisciplinary programs that showcase the GSMT as a preeminent cultural destination. They introduce diverse subjects as well as presentations on future developments in New York City by the building and construction industry.

As a case in point, on May 7, 2014, a significant audience turned out to hear the presentation *Hudson Yards: Building a New City within a City* given by Peter Sukalo, senior vice president of the Tutor Perini Corporation, who explored the largest real estate project in United States history. This monumental project will engage the services of numerous departments in the construction industry and create a ready-made neighborhood spanning twenty-six acres over what is now the rail yard on Manhattan's far West Side. The project will include ten million square feet of offices, apartments and luxury shopping malls where some sixty thousand people will live or work surrounded by New York's latest recreational destination, the High Line.

The Labor, Literature and Landmarks Lecture Series continues its presentation of *Behind the Scenes Of…*, featuring prominent Broadway production designers including Peter Nigrini, theatre projection designer; Abe Jacob, sound designer; and Jeff Davis, lighting designer.

WOMEN IN THE SOCIETY

With groundbreaking strides, the GSMT has kept pace—and in some cases, was ahead of the times—by admitting women into The Society's school in 1826 and the Apprentices' Library in 1862.

Today, The Society is embarking on a new era and welcomes women to join its roster of distinguished members. The first woman to sign on was Society member (1997) Pamela Jerome, an architect and conservator.

The GSMT continues to be on the cusp of modernity, and women have come to the forefront in the executive administration of the organization. For instance, Society member (2005) Victoria A. Dengel is the first woman to be elected president of The Society and served in this position in 2011 and 2012.

Dengel's dedication to The Society perhaps has strong roots in the fact that her father, Gerard A. Dengel, was president of The Society in 1982, and no doubt, The General Society was a frequent family topic that fueled her interest.

Dengel's consistent participation in the organization and her firm belief in the preservation of The Society led her on a new path of responsibility. In 2013, following her two-year tour as president of The Society, she was recruited by the GSMT Board of Governors to fill the position of executive director of the GSMT, a post that she holds today. Among her diverse responsibilities, Dengel oversees all the day-to-day operations of The General Society's headquarters—all this while keeping up-to-date with the building and construction industry and attending to the introduction of innovative programming. In 2013, Dengel also undertook the full responsibility for directing the Mechanics Institute.

Dengel's days are full and engaging, as she is actively involved with the multitasking role of administrator, and as such she sits on each of the The Society's committees whose members, in the spirit of brotherhood, volunteer their services just as the founding Brothers did before them. In 2015, the GSMT was recognized for industry education.

> *The General Society of Mechanics and Tradesmen was recognized for their dedication to "Industry Education" by The New York Association of Realty Managers, NYARM, January 24, 2015. Executive director and past president, Victoria A. Dengel accepted the award on behalf of the GSMT.*

Members Remembered

The Society's membership is a remarkable record that attests to the continued support of individuals in the building and construction trades to which the GSMT is so closely aligned. For instance, the GSMT's membership increased from 184 members in 1792, the year of The Society's charter, to nearly 4,500 members by 2014 as listed in The Society's register book. This rare book is of great historical value; it makes the past come alive as it records the numbered names and signature of every member who has joined The Society since its founding. Among the thousands of signatures are recorded the names of the founding members of The Society, whose integrity and foresight contributed to rebuilding a devastated New York City after the American Revolution. Their fortitude helped to create the strong country we have today.

The scuffed, brown leather binding of The General Society's original register book is no longer pristine but is historically preserved at The General

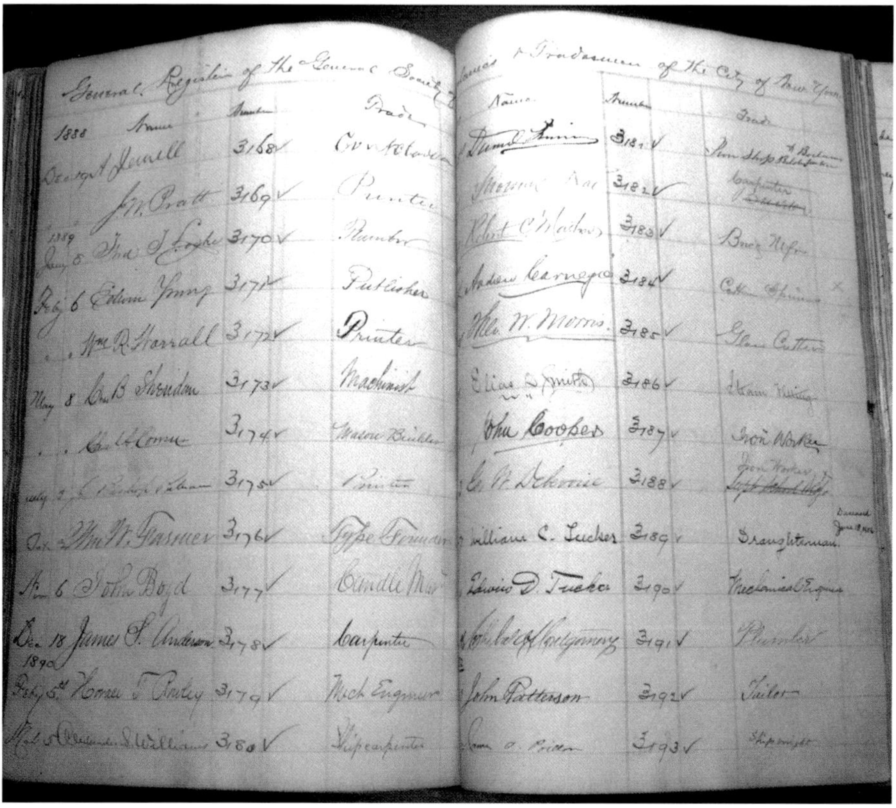

GSMT's historic registry book contains members' signatures from 1785 to the present. Note Andrew Carnegie's signature, number 3184, on the right-hand page.

Society. It records the signature of first chairman, Society member (1785) Robert Boyd, who served as chairman 1785 to 1786, and the signature of Society member (1786) John Campbell, a former general in the Continental army who was among the founders of The Society and chairman (1787–88). Campbell, as noted earlier in this book, was the grandfather of Peter Cooper, founder of The Cooper Union, who became a Society member and signed the register book in 1837. Another scion of the Cooper dynasty, Peter Cooper's grandson, the inventor Peter Cooper Hewitt, was also a member of The Society (1903). Peter Cooper Hewitt's inventive output was prolific, and working in collaboration with Frances B. Crocker, the origin of the modern-day helicopter for the United States Army began in their workshop.

Too numerous are the Brothers that deserve mention, so I reserve space to name but a few: the philanthropist Andrew Carnegie; John McComb Jr., the builder of the first city hall; Cornelius H. DeLamater, who built battleships for the Civil War; and George E. Hoe, who established prize awards for the Mechanics Institute graduates.

Throughout its history, The Society's committees have been the backbone of the GSMT's operations. Members, on a volunteer basis, working in tandem on a committee operate pretty much in the same manner as they have since the GSMT's founding. The chair of each committee is an active member of the Board of Governors, heading up such divisions as the library, the school, facilities, finance, development and good and welfare.

When the Facilities Committee, responsible for all the operations and maintenance of The Society's historic building, held their monthly meeting, preservation of the building itself to serve future generations was a major topic. With the chair, Rella Fogliano, the members assembled with Brothers Gottfried Weissgerber, Andrew Schon, Walter Bishop, Gerard Dengel Jr. and Joseph Humann to discuss the progress of the work being done on the building's façade, safety issues, maintenance issues in the building and the restored marble work. Most pertinent to the discussion was a guide from Easton Architects concerning the work needed to restore the building so it can continue to be part of the cultural heritage of the city.

INDEPENDENCE DAY CELEBRATION

Americans annually celebrate, not the Revolutionary War, but rather the Declaration of Independence. Since the founding of the Republic, The General Society of Mechanics and Tradesmen of the City of New York has been among the leaders in not only affirming but also in proclaiming the truths that the Declaration of Independence describes as "self evident."

The General Society, since 1793, has been commemorating the signing of the Declaration of Independence by listening to the words of that document that declared our independence as a free nation. The following excerpts from the GSMT annals provide a record of the proceedings of July 4, 1793:

> *The Declaration of Independence was read, after which a procession was formed, which proceeded to the old Presbyterian Church, where a sermon well adapted to the occasion was delivered by Rev. William Miller; when The Society dispersed, and again met at six o'clock P.M. at the house of Mrs. Amory, which was illuminated for the occasion. Deputations were*

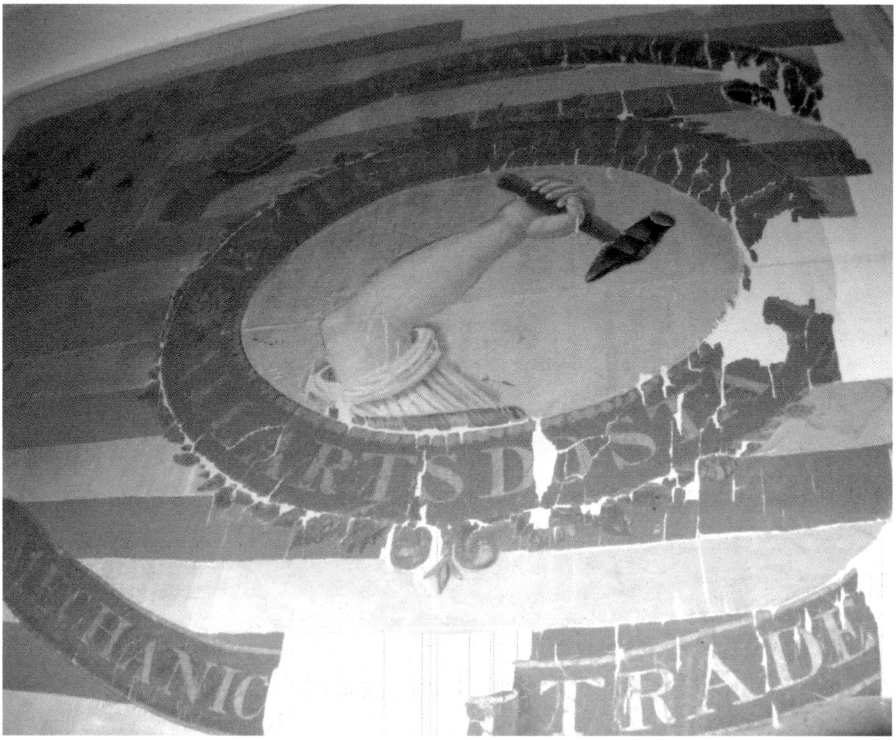

The torn and tattered original standard of The Society first unfurled at its anniversary festival on January 7, 1795, now preserved in The General Society's museum.

received from Tammany Society, The Society of Cincinnati, officers of the militia, and merchants of this city—a happy gathering congratulating each other on the return of the auspicious day. The evening was spent in the enjoyment of the entertainment and social intercourse which ever characterize this Society.

The patriotic observance and the reading of the Declaration of Independence are recorded again in The Society's minutes of 1795 with reference to the American flag.

In the year 1795, the American ensign, as it is called in the minutes, and as it is still exhibited by The Society on all public occasions, was adopted. At the meeting held July 1st, 1795, it was voted that this flag be displayed at sunrise. At eight o'clock A.M. The Society met to listen to the Declaration of Independence. The members then proceeded to join the general procession of citizens. At five o'clock P.M. The Society met again to receive the congratulations of the sister societies, and the citizens. Regular toasts of the usual character were given at the dinner. Among these, the following may be selected:

Franklin having died, April 17, 1790…
"The memory of our late brother mechanic, Benjamin Franklin: May his bright example convince mankind that in this land of freedom and equality, talents, joined to frugality and virtue, may justly aspire to the first offices of government." The processions formed at the Battery at nine o'clock A.M. and moved in the following order:

First.	*The Legion.*
Second.	*The General Society of Mechanics and Tradesmen.*
Third.	*The Tammany Society.*
Fourth.	*The Democratic Society.*
Fifth.	*The Militia Officers.*

The route of the procession was through Beaver, Broad, Pearl, and Beekman streets to the Presbyterian Church. Here the Declaration of Independence was read by Edward Livingston, Esq., and a sermon preached by Rev. Mr. Miller. A suitable ode was sung, and a collection taken up for the benefit of the charity school connected to the church. Afterwards the procession marched down Broadway to the Battery, where a feu de joie closed the ceremonies of the occasion.

The Society's annual Independence Day celebration, a continuous event since its inception over two hundred years ago, holds special significance for New Yorkers. They are not only reminded of the patriotic events of the past, but such patriotic observances bring to light current episodes concerning the American flag and 9/11.

For instance, of patriotic significance is the tattered "last flag flying" over the ruins of the World Trade Center as the final steelwork came down. On the road to recovery seven years after the planes struck the towers, the flag was carried on a bus to a tornado site in Greensburg, Kansas, where tornado survivors stitched it back together.

Since then, the National 9/11 Flag has traveled to fifty states where threads from other historic flags were added to this banner, such as bits from the original Star Spangled Banner, the flag that cushioned President

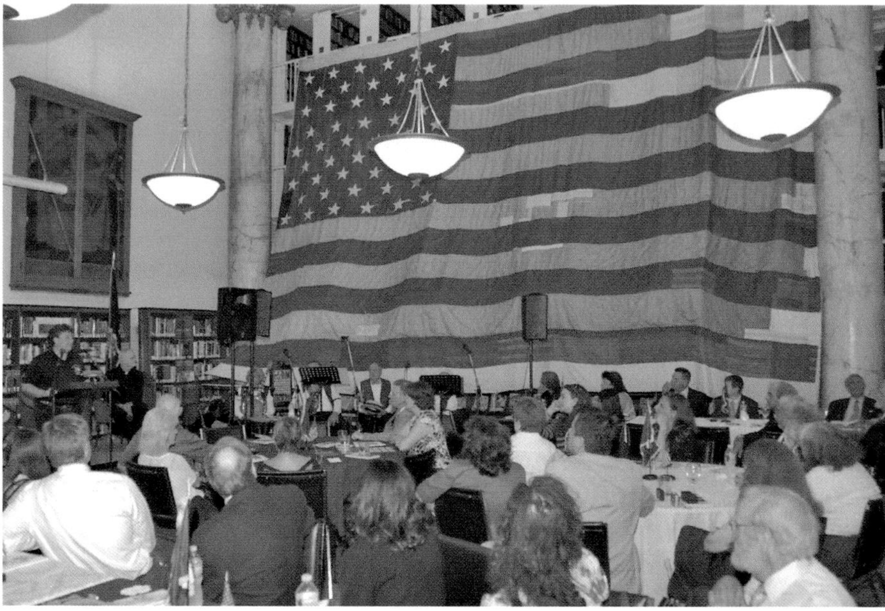

The colossal 9/11 flag in its temporary home at the GSMT. From the Independence Day celebration, 2012. *Photograph by D.A. Luhmann.*

Abraham Lincoln's head after he was shot and fragments of the celebrated flag left flying after Pearl Harbor. Afterward, servicemen from every state added their stitches.

The GSMT was privileged to have this huge twenty- by thirty-foot flag, which was so lovingly restored, hanging from the Library's three-story stacks in 2012 at a ceremony commemorating the fateful day. The flag was later delivered to the 9/11 Museum.

Since the GSMT's founding, the Independence Day celebration has been a highly coveted annual event, and in the spirit of patriotism, it continues to be observed at the GSMT headquarters today with a full reading of the Declaration of Independence. In 2014, the Fourth of July celebration marked the 238th birthday of the United States of America, and the GSMT continued its recognition of the great day.

By observing the historical past, we can appreciate what those before us endured to set us free. At the GSMT's annual celebration on July 2, 2014, held in The Society's Library, president Gottfried Weissgerber set the tone of celebration and welcomed all the distinguished guests, including past FDNY Commissioner Sal Cassano.

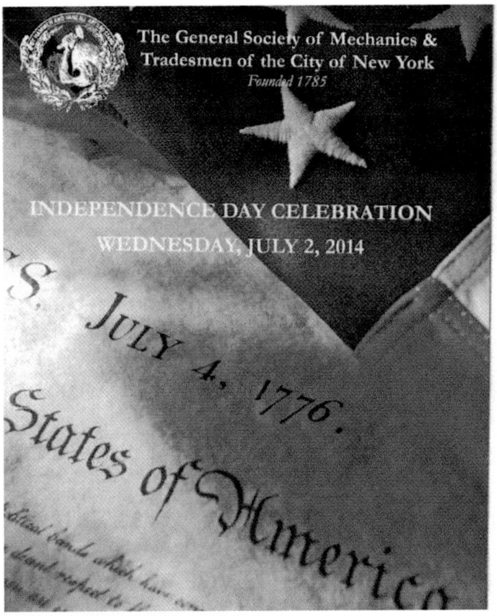

Independence Day Celebration
July 2, 2014

PROGRAM

7:00 P.M.
Reception .... Music by The Banjo Rascals

7:30 P.M.
Welcome and Salute to the Flag .... Gottfried Weissgerber
President, The General Society of Mechanics & Tradesmen

Star Spangled Banner .... Michelle DellaFave

Introduction of Patriotic Speaker .... Gottfried Weissgerber

Patriotic Address .... Terrance C. Holliday
Commissioner of the
Mayor's Office of Veterans' Affairs

God Bless America .... Michelle DellaFave

Musical Selections .... The Banjo Rascals

Reading of the Declaration of Independence .... Catherine Chadwick

America, the Beautiful .... The Banjo Rascals

Dinner with musical interludes with The Banjo Rascals

The Declaration of Independence annual celebration program held in the GSMT Library, July 2, 2014.

Michelle DellaFave rose to the occasion and sang the "Star-Spangled Banner," the song attributed to Maryland native Francis Scott Key, a thirty-five-year-old lawyer. This soupçon of 1812 history reminds us of the patriotism of our ancestors:

> *During the War of 1812 and the defense of Fort McHenry, located at the narrow entrance to Baltimore's harbor, Key witnessed the massive display of firepower, during the British assault on Baltimore, from the deck of an American flag-of-truce vessel. He had just completed negotiations with the British for the release of an American prisoner confined aboard a ship in the British fleet. On the morning of September 14th, Key realized that the British bombardment had been a failure when he saw the British squadron withdrawing downstream. The garrison flag was run up over the fort, confirming the post was still in American hands.*
>
> *Key was so moved by the sight of the flag defiantly flying over Fort McHenry that he wrote "The Defense of Fort McHenry." Later the poem was set to the 1770s old English song, "To Anacreon in Heaven" by John Stafford Smith. Do keep in mind, that the melody had been already a popular tune in Colonial America. Yet legend indicates that the tune was originally composed for the Anacreontic Society, ironically a British music club that held its meetings at the Crown and Anchor Tavern in London, England.*

However, not until 1931 was "The Star-Spangled Banner" declared, by an Act of Congress, the nation's official anthem. The tattered flag shot through and mutilated (original size; 40ft. x 32 ft.) that flew over Fort McHenry resides in the Smithsonian Museum. Of special note, a rare first edition of "The Star Spangled Banner," marking the 200th anniversary of the original famous anthem, is housed at The Morgan Library & Museum. It was on display to coincide with the patriotic holiday.

To continue comment on the GSMT's Independence Day celebration, New York City's former commissioner of the mayor's Office of Veterans' Affairs, Society member (2014), Terrance Craig Holliday, delivered the patriotic address and reviewed the GSMT's historical past and present significance. He congratulated The General Society for its service to the community in providing tuition-free classes at the Mechanics Institute.

The Declaration of Independence was read by GSMT friend and Library volunteer, the actress Catherine Chadwick, with all the poignant nuances the great work deserved.

DellaFave returned to the platform to sing "God Bless America," and the 2014 celebration ended with the customary conviviality: a celebratory dinner along with rousing patriotic songs, including "America the Beautiful" provided by the Banjo Rascals.

PATRIOTIC CELEBRATIONS

The annual Veterans Day holiday is yet another event of patriotic significance upheld by the GSMT since 2012. On this occasion, The Society hosted the weeklong events that began with an evening of rousing songs from all the military services. Visitors to the veterans' annual art show found on display an exhibition of memorabilia from all the branches of the military.

A highlight of the festivities at a recent Veterans Day celebration was welcoming five Congressional Medal of Honor honorees to the event. At that time, an impressive book, *Medal of Honor: Portraits of Valor Beyond the Call of Duty*, was presented to the GSMT by the author, Peter Collier. The book features portraits of valor from World War II, the Korean War, the Vietnam War, the Gulf Wars and the war in Afghanistan. Each medal recipient's story is different, inspiring and glorious.

The Medal of Honor originated in 1861 when President Abraham Lincoln began acknowledging individuals "for their gallantry and intrepidity in action at the risk of life above and beyond the call of duty." According to Barack Obama's letter from the White House, the text reads, "Fewer than 3,500 brave patriots have received the medal."

Flag Day, another long-standing patriotic event, is held on June 14 when the nation at large celebrates the anniversary of the adoption in 1777 of the Stars and Stripes by the Second Continental Congress. As early as 1793, the GSMT formed a contingent of Brothers who started this observation; its participation continued throughout the organization's history. Today, with pride and patriotic zeal, The General Society continues to be represented in New York City's Flag Day parade, carrying our country's colors and The Society's flag.

In 2014, among the flag bearers were the GSMT's executive director Victoria A. Dengel, GSMT members Bill Gonzalez (2004) and Erwin Schaub (2003) and Mechanics Institute registrar Meg Stanton. They all marched from the steps of city hall to Fraunces Tavern where George Washington said farewell to his officers at the close of the American Revolution.

Another GSMT historical record illustrates how much America's history is entwined with The Society's history and the Civil War's New York Volunteer Engineer Regiment:

> *The Society presented a Stand of Colors to the first New York Volunteer Engineer Regiment (this was the first regiment of its kind to be equipped and organized in the United States Service) which was composed exclusively of mechanics, many of whom were members of The Society. In 1868, the flags were returned to The Society after the disbanding of the regiment at the close of the Civil War and are now preserved in glass cases on the rear wall of the Library.*

THE MUSEUM

Throughout The Society's history, its members have made significant contributions, which have become a trove of historic documents in the GSMT's museum holdings. Society member (1893) William J. Hoe presented eighteen framed photographs of various sections of the city, which has changed materially since the pictures were taken; these images illustrate the transformation that is constantly going on.

Over the years the Brothers have stepped forward with significant personal documents. Society member (1908) James H. Schmelzel presented an original certificate of membership, dated February 7, 1838, issued to his father George J. Schmelzel. Society member (1839) John H. McCullagh presented an old certificate of appointment, dated July 13, 1812, of his grandfather as a fireman of the City of New York. These artifacts, which continue to be donated, chronicle the legacy of some of the Brothers' heritage.

For instance, the museum collection also includes original manuscripts, medallions, coins, paper money, books and curios of various kinds dating from antiquity up to the present. Of particular interest is a fragment of the Atlantic cable, which made possible the first transatlantic communication between New York and England. Brother (1888) Henry W. Merrill presented it to The Society.

Cyrus West Field was the originator of the Atlantic cable project, but it was largely Society member (1837) Peter Cooper, president of the New York, Newfoundland & London Telegraph Company, who financed the project during the years of experimentation. After the cable was successfully laid across the Atlantic Ocean in 1866 it shortened communication between the continents from several weeks to minutes. Today this artifact is displayed in The Society's executive office.

Among the documents of historical interest housed in the museum is a letter signed by President Abraham Lincoln dealing with an official matter and a letter of a personal nature signed by Benjamin Franklin.

Some of the museum's great treasures include a complete set of the *Scientific American* from its founding to the mid-twentieth century, to books and scientific reports concerning the building of the infrastructure of urban America.

When the Revolutionary War patriot Marquis de Lafayette was in New York as a guest of President George Washington, a letter was sent to Lafayette by The General Society proposing a visit. Though there is no record of such a visit taking place, the letter, as well as a medallion of the Marquis by sculptor David D'Angers, resides in the museum.

Lest we forget, one of the big historical events of 2015 takes us back to 1780, when a French frigate called the *Hermione* brought the Marquis de Lafayette to America with exciting news of renewed French aid for the faltering American Revolution. The "Frigate of Liberty" in which Lafayette sailed to America has historical significance in its role in turning the tide of the Revolutionary War. It therefore seemed an invaluable part of history not to be lost and the idea of reconstructing an authentic, historical replica of the *Hermione*, made by hand in the techniques of the eighteenth century, was conceived in 1993 by the organization Friends of Hermione-Lafayette in America.

Thomas Fleming, the eminent author and board member of the New York American Revolutionary Round Table, summed up what the *Hermione* means. He declared that without Lafayette, there probably would not have been a French alliance, and without French alliance, America would have lost its war of independence.

To commemorate that historic event, the completely recreated *Hermione*, a work of pure craftsmanship in glorious original colors, reminds us of the GSMT motto "By Hammer and Hand All Arts Do Stand." The *Hermione* sailed across the Atlantic in April 2015, visiting ports along the east coast of the United States for a summer of celebrations.

Artifacts in the GSMT

The second-floor executive office holds another treasure that honors one of the GSMT's great philanthropists. It is an imposing bronze bust of Brother Andrew Carnegie, signed in 1908 by the Scottish-American sculptor John Massey Rhind. The artist had immigrated to the United States in 1889 and settled in New York City and was a compatriot of the philanthropist. This perhaps explains the commission to this particular artist.

The bust was unveiled at a meeting of The Society in 1908 at a time when Carnegie had donated an additional $200,000. In making the donation, Carnegie said:

> *It is a wonderful institution, doing more good with a small sum of money than any I know of.*

The marble pedestal accompanying the statue was donated by GSMT member (1894) Alexander Walker, who was president in 1912. The bust, which is an excellent likeness of Carnegie, takes an honorary place at the entrance to the executive offices in recognition of Carnegie's benefactions.

Another bust, a plaster cast with a bronze-like finish, honors Louis Rouillion, a revered director of the Mechanics Institute (1900–45). He was an educator with a great vision, and legions of graduates are a monument to his memory.

He wrote several books, including *A Course in Mechanical Drawing*, and was the founder of the GSMT Alumni Association.

In the lobby of the GSMT building, a gripping artifact that commands attention is a monumental bronze sculpture of the Hammer and Hand, which was first installed and displayed above the door at The General Society's Mechanics National Bank, founded in 1810 at 33 Wall Street. Today, this artifact resides mounted on a wall above The Society's marble staircase.

On any occasion, should you have the opportunity to enter the institute's lobby, there are several bronze plaques that are worthy of your attention, including the bronze tablet in the lobby that commemorates the 200th anniversary of the birth of George Washington.

Then, too, there is Amos F. Eno, who made a bequest of the princely gift of $1.8 million to the GSMT. He also gets his due recognition in a bronze plaque. On November 7, 1915, the *New York Times* reported:

> *The gift was not so much in the nature of a surprise, because Mr. Eno had often expressed a great desire to become a member of the organization, and his disappointment was great at not being qualified for election, never having had any mechanical training, The president, Charles J. Richter said, "It is too early to formulate any definite plan for the use of the money, but it was intimated that the future policy of The Society will undoubtedly be enlarged in many beneficial ways."*

Library visitors may see an eye-catching large plaque, bright blue framed with gold leaves, which is installed high up on the wall at the far end of the Library. Among The Society's artifacts, such a plaque appears as an icon that was installed in 1802 on The Society's first building, the Mechanics Hall at 239 Broadway. The plaque also reappeared, in 1843, at the third location of The Society, its Crosby Street Annex at 472 Broadway.

THE EXECUTIVE OFFICES

The Society's executive offices house a copy of the Brooklyn Museum's original oil painting of Robert Boyd, a blacksmith and one of the twenty-two founders of the GSMT. Boyd was initiated into The Society in 1785, the year the organization was established, and is listed and signed as number six in the membership register book. He served as The Society's

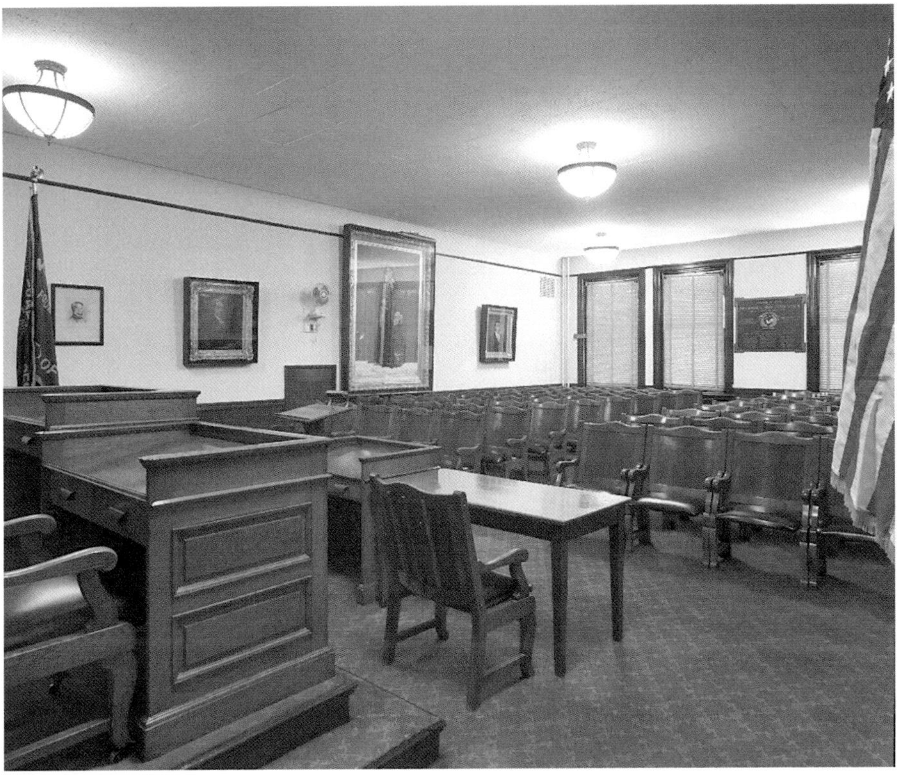

The GSMT's Assembly Room interior. A portrait of Thomas R. Mercein and the painting *From Riches to Poverty* hang on the sidewall. *Courtesy of Stephen Amiaga.*

first chairman in 1785 and 1786, and as the movement took hold, he steered the fledgling organization forward with exacting standards of high morals and honesty. He was The Society's first president in 1792, the same year of the GSMT's incorporation.

Boyd's portrait was donated by his descendant Society member (2013) George Boyd V and takes up its honorable place in the GSMT's executive offices.

The Assembly Room

A number of large nineteenth-century oil paintings also hang in the Assembly Room, including a portrait of Society member (1803) Thomas R. Mercein. He was the force behind the founding of the Apprentices' Library, as it was called when it opened in 1820. Today the Library is a vital public source of research and study in the heart of New York City's Clubhouse Row on West Forty-fourth Street.

Other rare paintings include a large canvas of General George Washington's dramatic crossing called *Passage of the Delaware* (1819), attributed to the American painter Thomas Sully. The monumental canvas of this magnificent painting with gilt frame embellished with patriotic regalia is encased in glass on an Assembly Room wall.

On many occasions, relatives of members contributed to the Assembly Room's holdings. A large panoramic rendering of Washington's Mount Vernon Estate as it was in 1845, attributed to Benjamin Henry Latrobe, was commissioned by Mrs. William Anderson Waydell in 1896. It was perhaps in memory of her father, Society member (1857) John H. Waydell, who worked in the cooper trade and was president of The Society (1885). The Waydell's affiliation with the GSMT runs deeper into The Society's history. Mrs. Anderson Waydell's grandfather, John Waydell Sr., was also a member of The Society (1832). This magnificent painting was presented to The Society in 1907 by Society member (1892) James O. Poillon, a master shipwright.

Of special pride, also in the Assembly Room, is the huge seventy-two- by sixty-inch oil painting *From Riches to Poverty*, by American artist Abraham Archibald Anderson, which portrays a woman with a small child clinging to her skirt. This relates to one of the founding tenets of the GSMT, a mandate to care for the widows and orphans of deceased members. In 1820, this charge led to the establishment of an elementary school for members' children, thirty-five years before New York City's free public school system began. William H. Webb, the master shipbuilder and founder of Webb Institute, presented the painting to the GSMT in 1886.

HONORING THE ORIGINAL TWENTY-TWO

As a testament to its history, a commemorative tablet installed in the Assembly Room honors the twenty-two founding fathers of the GSMT in 1785. It is interesting to note that, according to The Society's archives, a point of contention was made in 1894 by Brother Abram Stevens Hewitt regarding his grandfather John Campbell's name having been omitted from the tablet:

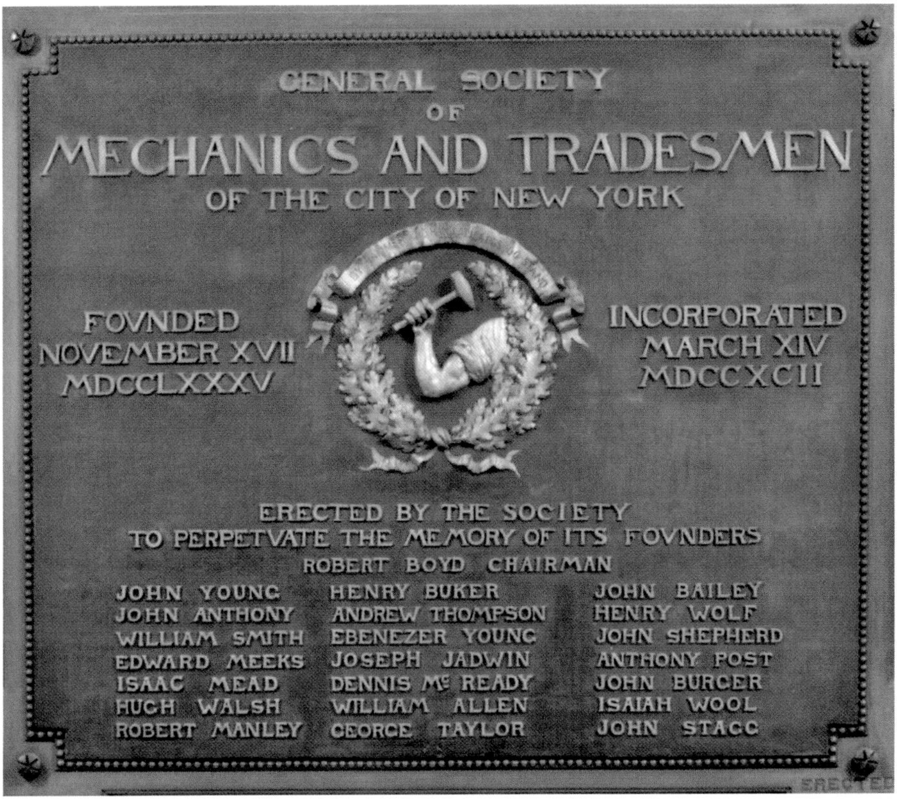

Bronze tablet memorial erected by The Society to perpetuate the memory of its twenty-two founders in 1785, now located in the Assembly Room. *Courtesy of Stephen Amiaga.*

> *Now, among the twenty-two men whose names are recorded upon the tablet which has been uncovered, I am sorry, not to see that of John Campbell, who certainly was an original member, and who became Chairman of the society immediately after Mr. Boyd; for there is nothing more treasured in the memories of Mr. Campbell's descendants, one of whom was Peter Cooper, than the fact that he was among the founders of this society.*

The omission of Campbell's name can best be explained that, although he was one of the early founders of The Society, he did not become a member until 1786, a year after the GSMT's founding and was chairman from 1787 to 1788.

Although the GSMT's bronze tablet memorial was erected to perpetuate the memory of its twenty-two founders, many other early members of The Society continued the organization's original goals and were given their due recognition in this tome. They, too, were visionary men who launched programs of benevolence, education and patriotism. The legions of men, the Brothers who followed in the footsteps of history, continued to be legend-makers of an organization that is in step with the introduction of modern and innovative programming that has influenced the educational and cultural development of the city.

THE JOHN M. MOSSMAN COLLECTION

In 1903, Society member (1874) John M. Mossman, the famed bank vault manufacturer, donated to The Society his singular collection of locks, keys and tools dating from 2000 BC. The curator of the Mossman collection, Society Member (2000) John Erroll, with his son Society member (2004) David Erroll, co-authored the book *American Genius, Nineteenth Century Bank Locks and Time Locks*. The book is a detailed and richly illustrated history of American bank vault locking technology in the nineteenth century as embodied in the John M. Mossman Lock Collection. Many of the handcrafted masterpieces were elaborately decorated with intricate engravings and colorful designs.

John Erroll said that Mossman was "a true Horatio Alger story" of rags to riches. Mossman built bank vaults at the New York Stock Exchange, the Manhattan Trust Company, the Bank of America and the Chemical National Bank, to name a few. However, with the United States' entry in World War I, safes and locks were often melted down; as a result, the Mossman collection is home to many one-of-a-kind masterpieces.

Earlier, in 1928, the GSMT published a book *The Lure of the Lock*, by Albert A. Hopkins. The book includes some rather interesting tales about masked burglars, inside jobs, powder, dynamite and the acetylene torch. The book also discusses famous robberies throughout the country. These tales obviously stimulated the market for high forbidding safes. The chapter "Memoirs of Mr. Mossman" recounts:

> *Mr. Mossman's business being bank locks, especially time locks, it came to his hand frequently to replace old style locks with later ones. In that way he accumulated a large collection of obsolete although highly inventive locks; locks made to order—not in commercial quantities. Mr. Mossman was so interested in his profession that he early began making a collection of locks*

and keys from all over the world. When the collection became of great value and importance Mossman presented it to The General Society.

Recently, a large safe was revealed when boxed archival records that had been stacked in front of it were removed to The Society's rare book room. The exquisite gilt bronze medallion on the top of the safe reads "Silas C. Herring patent," indicating that it was built to the highest standards of the time.

Herring joined The General Society in 1848 and engaged in a variety of business enterprises before establishing his safe company. Herring's safe is one of three mid-nineteenth-century safes that form an important part of the Mossman Lock Collection.

The historic Mossman collection is on display on the balcony of the Library. It is open to lock scholars and the general public. From time to time, new contributions to the Mossman collection add to its historic value. For example, not so long ago, the GSMT newsletter reported:

After attending a meeting of the American Locksmiths Association in Kentucky, several members of its Italian counterpart ERSI, (ERSI=Esperti Referme e Serrature Italia=Experts of Locks in Italy) stopped in New York to view the John M. Mossman Lock Museum. Associate Curator, Society member (2009) Joseph G. Reinis and GSMT's executive director Victoria A. Dengel hosted the evening. The ERSI members presented the GSMT with two publications of ERSI: The Cartouches of St. Eligius, *the patron saint of locksmiths, and the recently published* Statutes of the Ancient and Noble Guild of Blacksmiths, *the rules and regulations of locksmiths and metalworkers guild in Italy in 1690.*

The John M. Mossman collection continues in its role as the depository of many of the most unique locks, keys, tools and other paraphernalia associated with the bank vault business and is the only museum of its kind in the world.

Rare Book Treasures

To have a book that makes the past come alive is one of the rare treasures in the museum, as it reveals how The General Society's members have shown great diligence in the record keeping of their meetings. The minutes and

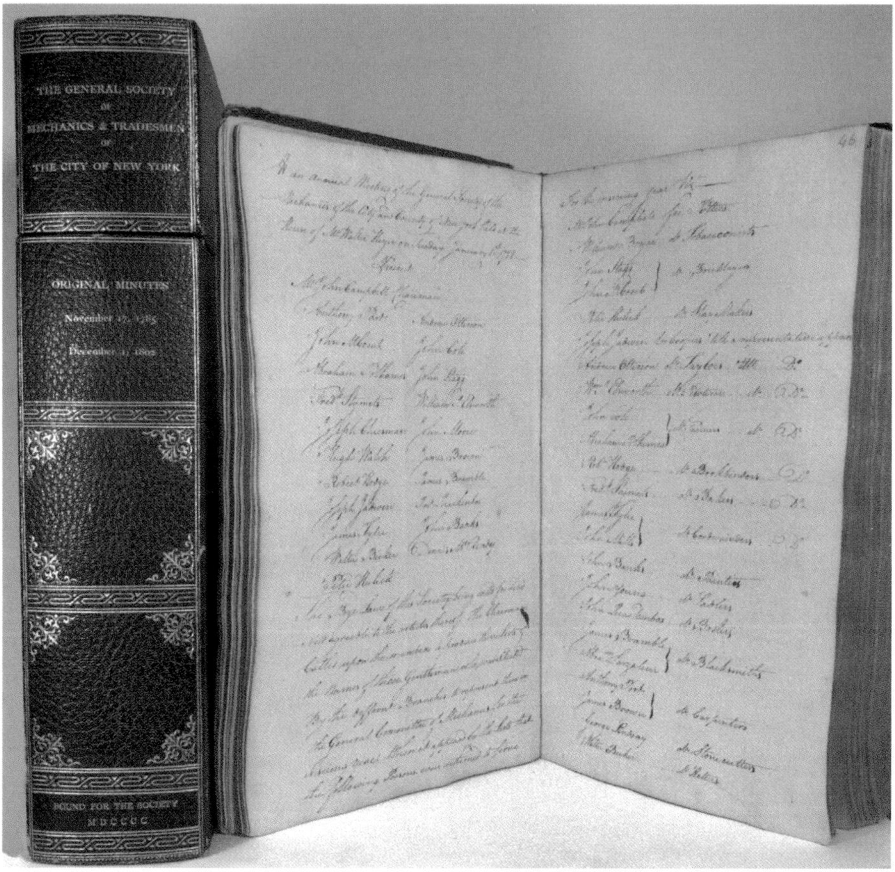

The original bound, handwritten minutes book, dating back to the GSMT's founding in 1785, in its own leather case with gold letters. It is a historical treasure in the rare book room.

resolutions were either written by hand or later typewritten and recorded in a bound book. One of the most interesting of these historical items is the original bound, handwritten minutes book, dating back to The General Society's founding in 1785:

> *It has been carefully housed in the GSMT rare book room in its own black leather slip-case with gold letters identifying the contents. It is shelved alongside a boxed; typed copy of the original minutes probably produced about 1885. The typed version has been pored over by researchers and scholars over the years. General Society members have been diligent in preserving the documents relating to its history, minutes of individual*

committees as well as minutes of member meetings. Another treasure, the original 1785 member register is still in use and new members each sign on a numbers line.

The GSMT's professional standards of record keeping, established from its founding, continue today. Committee meetings are held on a regular basis, and minutes are recorded and reports given at the monthly general meeting in the Assembly Room.

The financial security of the GSMT has had at its base member support and endowments from individuals or organizations. To this end, members of The Society, past and present, have bequeathed trust funds or provided legacies in their wills. Throughout their tenure, members have also contributed money designating its purpose for the Mechanics Institute or a specific project, such as new lighting in the Library. Such contributions throughout The Society's history have substantially enabled the work of the GSMT to be continued. This financial support by its members, as well as industry contributions, is a tradition that persists today.

Culture and Programs at the GSMT

Throughout the long existence of the GSMT, the Brothers made wise decisions to rent unoccupied space in their buildings to various tenants to garner income for The Society's school and benevolent activities. Such was the case in 1848 when at a meeting:

There was much discussion about this time concerning the propriety of leasing the lecture room to the Christy's Minstrels. The lease, however, was continued for a number of years and was a constant source of discussion and disagreement, many members regarding the entertainments of the Christy Minstrels with disfavor.

However, the Christys prevailed and stayed on as tenants for several years, thus providing income for the fledgling society. The GSMT has always kept pace, and very often, it has been ahead of the times with the practical application of economy. Its historical views are reflected in today's policy of renting space to cultural organizations or those in the creative industries who use the more spacious Library venue at the GSMT's 20 West Forty-fourth

Street headquarters for performances and lectures. To this end, the building's library space has become the site destination for numerous television, film, fashion shows and filmings.

For instance, a recent fashion show took place in this venue. It was the fashion designer Kenzo's pre-fall 2014 collection of women's wear. The models wearing the designer's new creations were stationed in front of the Library's three-story stacks, from where they walked down into the Library and formed a tableau at the front of the room before an audience of fashion insiders and the press. One of Kenzo's designers, Humberto Leon, commented that:

> *He doubted that any attendees had ever before stepped into the three-story reading room of this 229-year-old vocational organization, even though it was located on a well-traveled street with many landmark buildings.*

On another occasion, the GSMT made cinematic history in 1993 when a scene from Woody Allen's movie *Bullets Over Broadway* was filmed in The Society's building. The musical by the same name, based on the screenplay by Woody Allen and Douglas McGrath, opened at the St. James Theatre in 2014.

Keeping up with the trend to extend its venues, the PBS Series *Live from the Artists Den* filmed in The Society's Library space was a riveting scene of high drama:

> *It was an evening like no other! The 50th episode of this 3-time Emmy nominee,* The Artists Den, *was filmed on December 10th, 2013 in The General Society Library. The four hundred guests in the balconies and on the library floor witnessed a spectacular performance by the popular Grammy-nominated band, Vampire Weekend. Our serene Library had never before been the scene of such flashing lights and rock-music sounds, and the effect was breathtaking. The series creator, Mark Lieberman, selects exciting venues in which to film the show and GSMT's site was the perfect fit.*

On this occasion, the GSMT's executive director, Victoria A. Dengel, when interviewed about the program at The General Society said, "It was a unique opportunity to introduce our activities to a national audience." The program was aired on PBS in late February 2014.

At a more recent presentation on October 28, 2014, master framers Eli Wilner and Company brought history and current events together in an Artisan Lecture

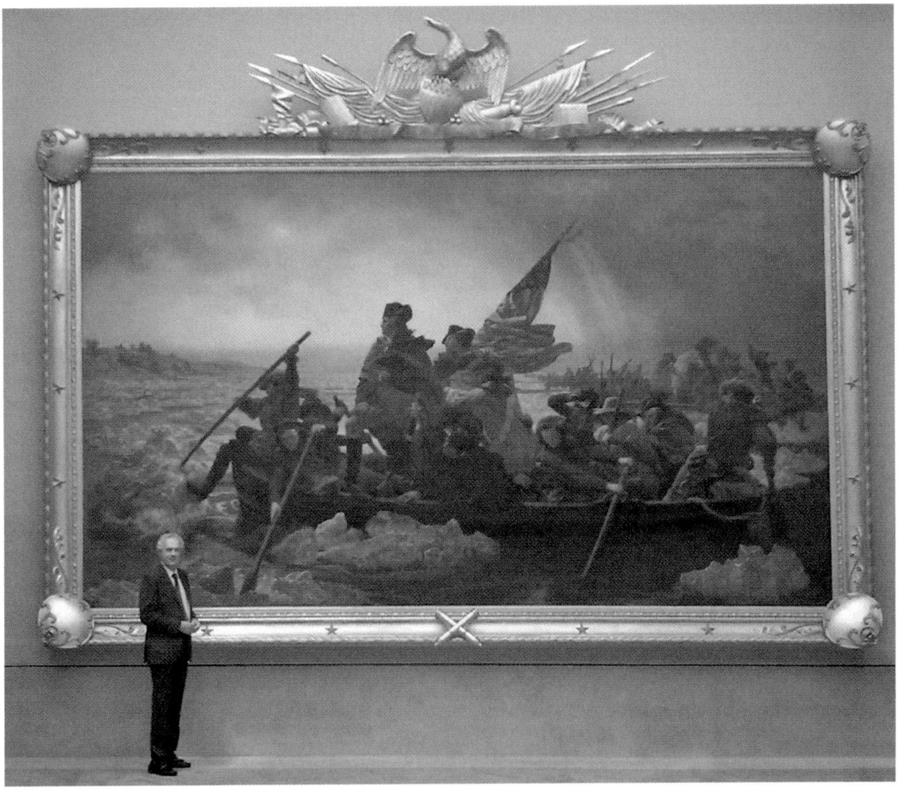

Washington Crossing the Delaware. The ornately carved gilt frame recreated by Eli Wilner and Company for the Metropolitan Museum of Art. *Courtesy of Eli Wilner, pictured at left.*

Series program presented by independent frame historian Suzanne E. Smeaton. The lecture was called "The Story of How an American Masterpiece Was Replicated." Smeaton discussed frame antiquity and Wilner's role as a master frame maker and conservator, particularly their artistic vision in creating the colossal hand-carved and gilded replica of the lost original frame for *Washington Crossing the Delaware* by Emanuel Leutze at the Metropolitan Museum of Art.

As the museum prepared to open its new American Wing, a Mathew Brady photograph from 1864 came to light, showing the twelve- by twenty-one-foot masterpiece in its original frame, which had been lost over one hundred years ago. Only one framer had the expertise, ability and facilities to recreate it, so the museum turned to Eli Wilner to help restore the frame's full visual depth and drama. Ornately carved, the awe-inspiring frame displays an eagle at the center surrounded by flags, pikes, a banner and other regalia.

It is significant to be reminded that the GSMT's motto "By Hammer and Hand All Arts Do Stand" represents the skilled artisans whose handcrafted work, in those early days of the new republic, often were made with tools that they created themselves. History repeats itself in the Wilner lecture, which pays homage to the old world skills of the craftsmen who used the same tools, techniques and materials as the originals to create the colossal frame.

PROUD OF ITS HERITAGE

The GSMT has kept up with modern trends, transformed its activities and responded to innovation. Today, The Society represents itself in its true character as a major educational and cultural center. It continues to serve the public through the Mechanics Institute, its extensive lecture series, patriotic ceremonies and cultural programs that have enriched the quality of life in New York City.

In summation, this statement extracted from the annals of the GSMT (1785 to 1880) still resonates with The Society's proud heritage. Society member (1871) John Banta, vice president (1880) and president (1881), reflected on The Society's first one hundred years:

> *For a period of little less than 100 years the GSMT has been doing the work contemplated at the time of its original institution. Public and private edifices give evidence of the skills and abilities of New York mechanics. Our mechanics sustain a reputation for integrity and fair dealing; our commercial transactions extend to every quarter of the globe; ere long New York may be found the business and money center of the world. Our membership includes the names of those who, in their day and generation, were foremost in the ranks of good and true men. May those who occupy their places continue the work that was begun by them and prove worthy of such an ancestry. May the future of The General Society of Mechanics and Tradesmen be worthy of its honorable past!*

From the GSMT annals:

> *The General Society of the Mechanics and Tradesmen of the City of New York is justly proud, not only of its origin, but its history when*

people—the Knickerbockers of Manhattan Island—were honest and domestic, industrious and patriotic, conservative and religious.

Now the GSMT enters one of the most exciting times in its cultural life, and at the same time, it maintains the integrity of its original goal: to serve the public through education, lectures and innovative venues that resonate with the cultural interests of the citizens of New York. The GSMT building itself is one of the most important sites in the city and stands as it did in 1899, firmly planted in the history of New York City. This book also gives you an insight into the rich archives preserved by the GSMT that made this book possible.

Looking ahead, The Society embraces the splendor of a new era that is witnessing the rebirth of the craft tradition, the way of making artifacts and products by hand. It continues to honor its motto "By Hammer and Hand All Arts Do Stand" and never forgets its origins, which rose in tandem with New York City's history to such prominence.

The GSMT also honors the building and construction industry, which is so brilliantly woven into the fabric of the GSMT's history. Then, too, it honors those hero laborers who continue to transform the skyline of New York City to such incredible heights.

With such an illustrious past, The Society was naturally desirous that all that it has accomplished should be made a matter of historic record in this monograph. Something of this is due to the memory of the excellent men who established The General Society of Mechanics and Tradesmen of the City of New York and the Brothers who followed and carried it through so many generations in usefulness and honor. While the GSMT reveres its ancestors, the organization advances forward in the footsteps of history and continues to progress with an innovative and inventive spirit, today and all the tomorrows to come.

EPILOGUE

The General Society of Mechanics and Tradesmen of the City of New York, with its long-standing reputation as an epicenter of culturally driven activities, continues to impact the creative global economy of the city. Everything that is old about the organization is fresh and exciting with innovative educational and philanthropic programs that resonate with modern New Yorkers. Putting a new spin on its insignia motto, represented above, the GSMT has modernized its design to upgrade its important identity with the Mechanics Institute.

Whether it is through lectures or cultural events, the GSMT keeps up its original tradition of public service with a broad range of projects that bring to light the diverse relationship the organization holds with the labor, literature, artisan, preservation and construction communities.

The GSMT's Mechanics Institute, always aware of the importance of new technology, introduced BIM and CAD, to the courses in this unique tuition-free technical school. It continues to count among its graduates employees in the building and construction trades who have upgraded their training to the next level of expertise.

With its epic motto, The Society also pays homage to its Labor, Literature and Landmarks Lecture Series, which continues a tradition of public lectures that started at The Society in 1837. The lecture series brings people of diverse interests from the entire New York area and international visitors to the organization's spacious Library reading room, which is also the site of numerous cultural events. The Library itself, with its extraordinary book collection that dates back to the 1700s, is yet another venue for study and research by membership subscription.

The Artisan Lecture Series offers lectures that pay tribute to the art of craftsmanship by featuring master artisans who lecture about the intricacies of their specialized craft. These lectures promote the work and art of skilled craftsman, thus ensuring that their unique knowledge is understood and carried forth for future generations.

The GSMT's tradition of observing patriotic holidays, which began in the early years of the Republic, is faithfully reenacted even today. A major high point in these celebrations is the reciting of the Declaration of Independence, which reminds New Yorkers of their heritage that is so closely linked to America's.

On the occasion of the 2014 Annual Craftsmanship and Civic Leadership Awards, the following proclamation from the Manhattan Borough president Gale A. Brewer, sums up the GSMT's vital role in New York City.

Office of the Manhattan Borough President
City of New York

Proclamation

Whereas: We are proud to acknowledge those organizations and individuals that have remained committed to improving our city and vibrantly enriching the community; and

Whereas: The General Society of Mechanics and Tradesmen of the City of New York has served and educated Manhattan residents since its founding by skilled craftsmen in 1785; and

Whereas: The Society's educational, philanthropic, and cultural programs provide vocational guidance and professional development opportunities to New Yorkers from all backgrounds; and

Whereas: The Society's Mechanics Institute—one of the city's first free schools—was founded in 1820 and continues to provide tuition-free evening instruction in trades-related content and a unique service to improve job opportunities for New Yorkers in the building and construction industry; and

Whereas: supporting The Society's educational programs, the Apprentices' Library—the city's second oldest library and one of its three remaining private-membership circulating libraries—makes its extensive collections available to other institutions and the public; and

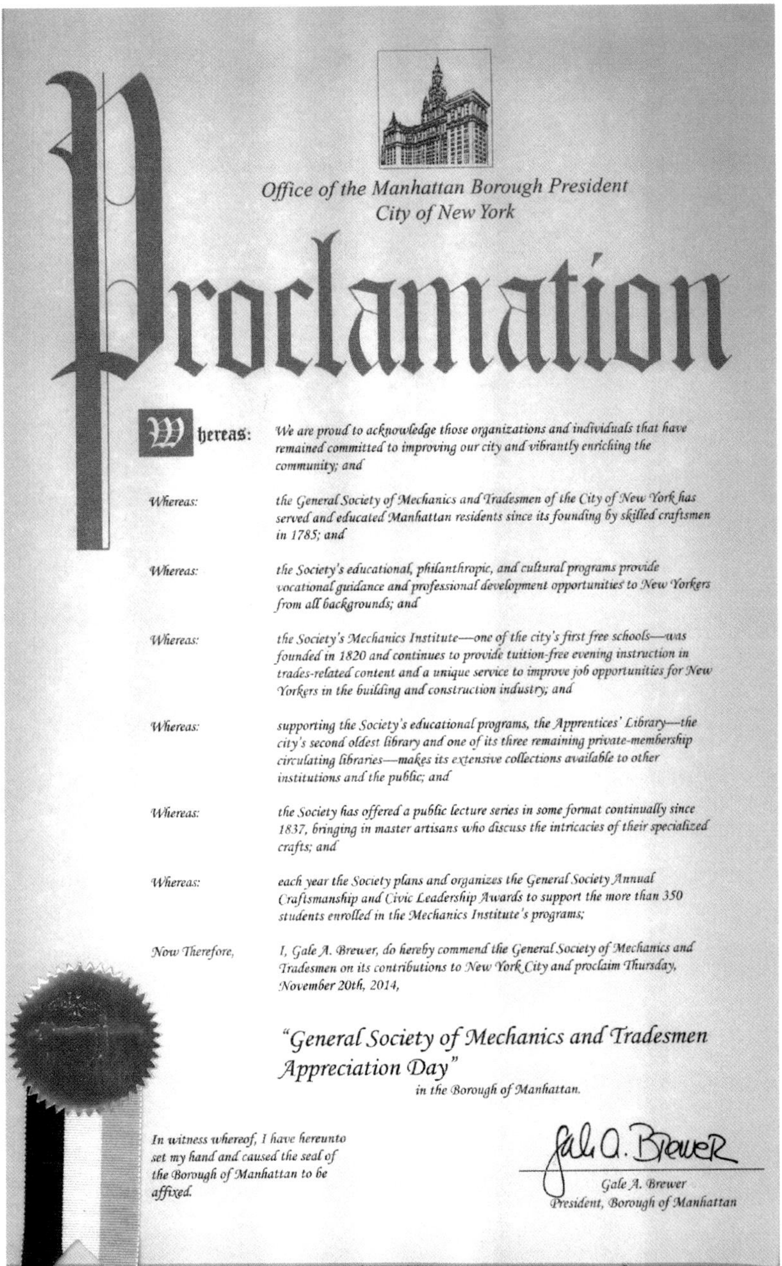

Office of the Manhattan Borough President
City of New York

Proclamation

Whereas: We are proud to acknowledge those organizations and individuals that have remained committed to improving our city and vibrantly enriching the community; and

Whereas: the General Society of Mechanics and Tradesmen of the City of New York has served and educated Manhattan residents since its founding by skilled craftsmen in 1785; and

Whereas: the Society's educational, philanthropic, and cultural programs provide vocational guidance and professional development opportunities to New Yorkers from all backgrounds; and

Whereas: the Society's Mechanics Institute—one of the city's first free schools—was founded in 1820 and continues to provide tuition-free evening instruction in trades-related content and a unique service to improve job opportunities for New Yorkers in the building and construction industry; and

Whereas: supporting the Society's educational programs, the Apprentices' Library—the city's second oldest library and one of its three remaining private-membership circulating libraries—makes its extensive collections available to other institutions and the public; and

Whereas: the Society has offered a public lecture series in some format continually since 1837, bringing in master artisans who discuss the intricacies of their specialized crafts; and

Whereas: each year the Society plans and organizes the General Society Annual Craftsmanship and Civic Leadership Awards to support the more than 350 students enrolled in the Mechanics Institute's programs;

Now Therefore, I, Gale A. Brewer, do hereby commend the General Society of Mechanics and Tradesmen on its contributions to New York City and proclaim Thursday, November 20th, 2014,

"General Society of Mechanics and Tradesmen Appreciation Day"
in the Borough of Manhattan.

In witness whereof, I have hereunto set my hand and caused the seal of the Borough of Manhattan to be affixed.

Gale A. Brewer
Gale A. Brewer
President, Borough of Manhattan

Manhattan Borough president Gale E. Brewer's proclamation presented to the GSMT on November 20, 2014, at the Annual Craftsmanship and Civic Leadership Awards.

Whereas: The Society has offered a public lecture series in some format continually since 1837, bringing in master artisans who discuss the intricacies of their specialized crafts; and

Whereas: each year The Society plans and organizes The General Society Annual Craftsmanship and Civil Leadership Awards to support the more than 350 students enrolled in the Mechanics Institute's programs;

Now Therefore, I, Gale A. Brewer, do hereby commend The General Society of Mechanics and Tradesmen on its contributions to New York City and proclaim Thursday, November 20, 2014,

"General Society of Mechanics and Tradesmen Appreciation Day" in the Borough of Manhattan.

In witness whereof, I have hereunto set my hand and caused the seal of the Borough of Manhattan to be affixed.

Gale A. Brewer
President, Borough of Manhattan

This 230-year-old organization, The General Society of Mechanics and Tradesmen of the City of New York, has a long and storied ancestry, but its ability to embrace change and remain innovative means it is still a vital part of the City in the twenty-first century. At every turn, it reinvents itself, looking to do more and do it even better. As such, The Society intensifies its mission to serve and improve the quality of life of the people of the city of New York.

APPENDIX 1

LANDMARKS AND RESOURCES

The General Society of Mechanics and Tradesmen of the City of New York, 20 West 44th Street, New York, NY 10036. (212.840.1840).

Alexander James Hill, Historian/Author, 154 East 89th Street, New York, NY 10028. (917.744.7184).

Amiaga Photographers Inc., 131 Sterling Place, Amityville, NY 11701. (631.943.6504).

Clermont State Historic Site, 1 Clermont Road, Germantown, NY 12526. (518.537.4240).

Cooper Hewitt, Smithsonian Design Museum, 2 East 91st Street, New York, NY 10128. (212.849.8400).

Eli Wilner & Company, 1525 York Avenue, New York, NY 10028. (212.744.6521).

EverGreene Architectural Arts, 450 W. 31st Street, 7th Floor, New York, NY 10001. (212.244.2800).

The FDNY George F. Mand Library, FDNY Fire Academy, Randall's Island, New York, NY 10035. (212.360.4413)

Fraunces Tavern, 54 Pearl Street, New York, NY (212.968. 1776).

John Jay Homestead State Historic Site, 400 Route 22, Katonah, NY 10536. (914. 323. 565l).

Marcus Robinson's Art Rep: Joyce Chasan, Artsource Int'l LLC., 333 Park Avenue South, New York, NY. (917.295.5016).

The Morgan Library & Museum, 225 Madison Avenue, New York, NY 10016. (212. 685.0008).

The New York American Revolutionary Round Table, NYARRT, c/o Chairman David Jacobs, djacobs01@snet.net.

New York City Fire Museum, 278 Spring Street, New York, NY 10013. (212.691.1303).

The New York Freedom Trail, http://nyfreedom.com.

PRATT Institute, 200 Willoughby Avenue, Brooklyn, NY 11205. (718.636.3600).

Silverstein Properties, Inc., 7 World Trade Center, 250 Greenwich Street, New York, NY 10007. (212.551.7362).

WEBB Institute, 298 Crescent Beach Road, Glen Cove, NY 11542-1398. (516.671.2213)

APPENDIX 2

Historic English Terminology Used in the Book

apprentice: One bound by legal agreement to work with another for a given length of time in return for instruction in a trade, art or business.

blacksmith: One who forges and shapes iron with an anvil and hammer.

coach builder: Designates a craftsman who specializes in building coaches.

cooper: One who makes or repairs wooden tubs and casks.

gilder: A person whose work is gold leafing.

Japanner: Works in Chinoiserie and varnish.

journeyman: One who has fully served an apprenticeship in a trade and is a qualified worker in another's employ.

master craftsman: A person who has achieved the highest rank, operates his own business and employs other workers.

master mason: An expert mason. One who has achieved the third degree of freemasonry.

millwright: A person who designs, builds or repairs mills or mill machinery.

nailers: Workers in the shipbuilding trade.

pewterer: One who makes pewter products.

quill maker: Creates writing pens made with a quill.

saddler: Specializes in saddle making.

sailmaker: Makes sails for oceangoing ships.

sawyer: One employed at sawing wood, as in a sawmill.

ship carpenter: A skilled worker employed in the shipbuilding trade.

ship chandler: A person who sells diverse equipment for ships including navigation devices, maps and telescopes.

shipwright: A skilled worker employed in the construction or maintenance of ships.

solderer: A person who joins metals together.

stonecutter: One who cuts or carves stone.

tallow chandler: Maker of candles.

watchmaker: Creates old-time watch pieces.

Bibliography

Books

Booth, Mary L. *The History of the City of New York.* New York: E.P. Dutton and Company, 1880.

Brown, Henry Collins. *The Story of Old New York.* Illustrated by Frank Rennie. New York: E.P. Dutton and Company, 1934.

Burrows, Edwin G., and Mike Wallace. *Gotham: A History of New York City to 1898.* New York: Oxford University Press, 2000.

Carnegie, Andrew. *Autobiography of Andrew Carnegie.* Boston, MA: Houghton, Mifflin, 1924.

Congdon, Charles T., ed. *Annals, 1785–1880.* Compiled by Thomas Earle. General Society of Mechanics and Tradesmen of the City of New York, 1882.

Ellis, Edward Robb. *The Epic of New York City.* New York: Kodansha USA, 1997. First published 1966 by Coward-McCann.

Erroll, John, and David Erroll. *American Genius: Nineteenth Century Bank Locks and Time Locks.* New York: Quantuck Press, 2006.

General Society of Mechanics and Tradesmen of the City of New York. *Annual Record Books,* 1829–1999.

———. *Historical Timeline,* 1787–1996.

———. "Mechanics Institute." Wikipedia. http://en.wikipedia.org/wiki/Mechanics_Institutes.

———. *Romance and Realism of a Great Achievement: Mechanics Institute, 1785–1932.*

Glynn, Tom. "Books for a Reformed Republic: The Apprentices' Library of New York City, 1820–1865." *Libraries and Culture* 34, no. 4 (1999).

Greene, Janet Wells. "The Library of The General Society of Mechanics & Tradesmen." In *America's Membership Libraries*. Richard Wendorf, ed. New Castle, DE: Oak Knoll Press, 2007.

Guérin, Polly. *The Cooper-Hewitt Dynasty of New York*. Charleston, SC: The History Press, 2012.

Historical Sketch and Government of The General Society of Tradesmen of the City of New York, 1785–1914. Cornell University Library Digital Collections, 1914.

Homberger, Eric. *The Historical Atlas of New York City: A Visual Celebration of 400 Years of New York City's History*. New York: Henry Holt and Company, 1999.

Hopkins, Albert A. *The Lure of the Lock*. New York: General Society of Mechanics and Tradesmen of the City of New York, 1928.

Hornung, Clarence P. *Treasury of American Design*. Vol. 1. New York: Harry N. Abrams, 1972.

Johnson, Herbert Alan. *John Jay, 1745–1829*. New York: University of the State of New York, 1975.

Johnston, Henry P. "Evacuation of New York by the British." *Harper's New Monthly Magazine*, November 1883, 909–23.

Kramer, Eric, and Carol Sletten. *New York Freedom Trail: A Guide to the City's Revolutionary War Sites*. New York: Wolf Water Press, 2014.

Lamb, Martha J. *History of the City of New York: Its Origins, Rise and Progress*. Vol. 11. Chicago, IL: A.S. Barnes and Company, 1880.

Martin, Lawrence. *Serving New York: 1785–1960, 175th Anniversary*. General Society of Mechanics and Tradesmen of the City of New York, n.d.

Milliken, Donald D. *Phyfe Duncan: Colliers Encyclopedia*. New York: Macmillan Educational Corporation, 1976.

Moscow, Henry. *The Street Book: An Encyclopedia of Manhattan's Street Names and Their Origins*. New York: Hagstrom Company, 1979.

Reed, Robert: *A Furniture Style that has Endured*. Discover Vintage America. http://discoverypub.com/feature/2011_01.html.

Riker, James. "Evacuation Day" 1783. Self-published, 1883.

Rock, Howard B. *Artisans of the New Republic: The Tradesmen of New York in the Age of Jefferson*. New York: New York University Press, 1984.

Speckter, Penny, ed. *GSMT News: The Newsletter of The General Society of Mechanics & Tradesmen*. 2000–2015.

Webster's Illustrated Encyclopedic Dictionary. New York: Houghton Mifflin, 1990.

NEWSPAPERS AND RESOURCES

AMNew York. http://www.amny.com. Valuable resource for Revolutionary War history.

Duncan Phyfe honored jointly by the GSMT and the Metropolitan Museum of Art in a lecture to celebrate the sesquicentennial anniversary of the GSMT, 1935. A letter in the Thomas J. Watson Library, Metropolitan Museum of Art, New York.

Fields, Sidney. "Only Human." *New York Daily Mirror*, April 1, 1957.

Kinnai, Clark, ed. "New York Queen of Commerce: A Pictorial History of the City's Trade." *New York Journal-American*, November 20, 1952.

McNeur, Catherine. "John McComb, Jr.: The Architect of the Federalist Party." New York University, December 2002.

New Yorker. "Profiles: Andrew Gunnar Hagstrom." April 9, 1949.

New York Evening Post. "Greater Demand for Books." November 27, 1820.

New York Sunday Times. "Centennial Celebration." September 6, 1885.

New York Times. "Abram Stevens Hewitt." January 7, 1900.

———. "Carnegie Says He's Glad to be a Poor Man's Son." November 19, 1905.

———. "$185,000 Spent in 1931 by Mechanics Society." March 12, 1932.

———. "$200,000 By Carnegie to Institute School." May 7, 1908.

———. "What Society of Mechanics and Tradesmen Is, Nearly Two Million by Mr. Eno." November 7, 1915.

New York World. "Dinner at Delmonico's." November 17, 1885.

———. "New York Celebrates Centennial." 1886.

Speckter, Penny. *The Philanthropic Activities of the GSMT, 1785–1885*. Master's thesis, 1986.

Swanwick, Morris. "The General Society of Mechanics and Tradesmen." *Music Trades*, December 2, 1905, 11–14.

Wyandt, Frieda. "Clean Lumber Made Dykes a Millionaire." *New York Daily Investment News*, July 25, 1929.

About the Author

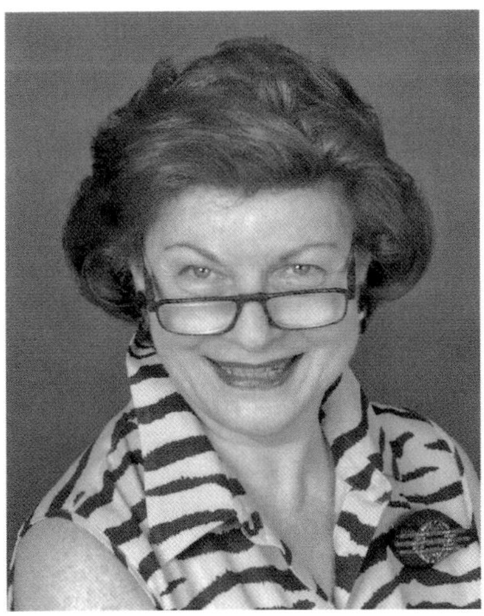

Polly Guérin is a former adjunct professor at the Fashion Institute of Technology in New York and the author of four college textbooks and two video productions. *The General Society of Mechanics and Tradesmen of the City of New York* is her second book for The History Press.

Guérin's features on the decorative arts, antiques, collectibles and design have appeared in *Art & Antiques* magazine and other consumer publications. She also writes on fashion and Art Deco and is currently the author of four blogs: "Pollytalk from New York," "Men Remarkable Visonaries," "Women Determined to Succeed" and "Poetry from the Heart by Polly Guérin."

Guérin counts among her professional memberships—and is a board member of—the American Revolutionary Round Table (ARRT) and the Giulio Gari Foundation, is a former board member of the Art Deco Society of New York (ADSNY) and now serves on its Advisory Panel. She is also a member of the Silurians, the Victorian, Browning and World Ship Societies and the Navy League. Visit her website at www.pollytalk.com.